The *Cosmos* as a Physics Organism

Stephen Blaha Ph. D.
Blaha Research

MMXXV

.

ISBN: 979-8-9986071-2-7

Rev. 00/00/01 July 5, 2025

To Margaret

Some Other Books by Stephen Blaha

SuperCivilizations: Civilizations as Superorganisms (McMann-Fisher Publishing, Auburn, NH, 2010)

Unification of God Theory and Unified SuperStandard Model THIRD EDITION (Pingree Hill Publishing, Auburn, NH, 2018).

The Exact QED Calculation of the Fine Structure Constant Implies ALL 4D Universes have the Same Physics/Life Prospects (Pingree Hill Publishing, Auburn, NH, 2019).

Passing Through Nature to Eternity ProtoCosmos, HyperCosmos, Unified SuperStandard Theory (Pingree Hill Publishing, Auburn, NH, 2022).

HyperCosmos Fractionation and Fundamental Reference Frame Based Unification: Particle Inner Space Basis of Parton and Dual Resonance Models (Pingree Hill Publishing, Auburn, NH, 2022).

The Cosmic Panorama: ProtoCosmos, HyperCosmos, Unified SuperStandard Theory (UST) Derivation (Pingree Hill Publishing, Auburn, NH, 2022).

God and and Cosmos Theory (Pingree Hill Publishing, Auburn, NH, 2023).

Newton's Apple is Now The Fermion (Pingree Hill Publishing, Auburn, NH, 2023).

Cosmos Theory: The Sub-Particle Gambol Model (Pingree Hill Publishing, Auburn, NH, 2023).

Cosmos-Universe-Particle-Gambol Theory (Pingree Hill Publishing, Auburn, NH, 2024).

Fractal Cosmos Curve: Tensor-based Cosmos Theory (Pingree Hill Publishing, Auburn, NH, 2024).

The Eternal Form of Cosmos Theory Third Edition (Pingree Hill Publishing, Auburn, NH, 2024).

Fundamental Constants of Cosmos Theory and The Standard Model (Pingree Hill Publishing, Auburn, NH, 2024).

Geometric Cosmos Geometric Universe (Pingree Hill Publishing, Auburn, NH, 2024).

Structure and Dynamics of Cosmos Theory and the Unified SuperStandard Theory (Pingree Hill Publishing, Auburn, NH, 2024).

Black Holes, White Holes, and Superluminal Starship (Pingree Hill Publishing, Auburn, NH, 2025).

The Sequence Theory Foundation of Cosmos Theory and Gambol Theory (Pingree Hill Publishing, Auburn, NH, 2025).

BEYOND TIME AND SPACE Fermion Strata & Consistency Conditions for Sequences, Cosmos Spaces and Gambols (Pingree Hill Publishing, Auburn, NH, 2025).

Available on Amazon.com, bn.com, Amazon.co.uk and other international web sites as well as at better bookstores.

CONTENTS

FIGURES and TABLES

Introduction

This book has a number of very significant features. Perhaps the most important new result is the reduction of the origin of the Cosmos to a single point that grows to the Cosmos spaces *in one step*. Chapter 22 shows the Master Equation solution embodies the entire evolutionary sequence of the Cosmos. Cosmos spaces flow from the Master Equation. This result eliminates the view that Cosmos features come from a Math grab bag. We see a closely crafted evolution from a point that is rather like the evolution of life. The growth of Cosmos Theory, Sequence Theory and Gambol Theory is intertwined. The total of these theories that follow has all the ramifications of a life form – complexity generated from what might be viewed as a seed of nothingness.

The theory now gives an alternative to having a simple set of driving forces for step by step evolution. We will see it encompasses the initial mathematical point, the generation of a two dimension piecewise linear unit length, the generation of the fractal Cosmos Curve, the generation of the Cosmos spaces and their dimension arrays, the unification spaces sequence, and the set of fermion mass sequences.

The Master Equation is the driving force of Physical evolution. Above all, it must be remembered that this evolution is instantaneous, and more importantly, beyond time, since time does not exist until universes are created.

The book has many points of interest:

1. The development of a quadplex theory of fermions and their interactions based on defining fermions to have four parts – some parts are bradyon (slower than light) – some parts are tachyon (faster than light). The theory nicely encompasses the fermion mass sequences. Four parts, four internal coordinate systems, and sequences of four fermions.
2. The Purpose of Physics.
3. Structure and Substance.
4. What is Mass-Energy?
5. The True Fundamental Element.
6. The basis of universes.
7. Detailed view of the fermion Strata and fermion Gambols of the theory including their spectrums.
8. Quad-Quad equations for ElectroWeak and Strong interactions.
9. Super-bradyon-tachyon (SBT) quantum field theory.
10. Gambol constituents of fermions.
11. Stratum and Gambol groups.
12. The spectrum of fermions and gambol fermions.

Highly recommended for a view looking to the future of Physics.

Some Fundamental Issues

1. The Emergence of Organic Life on Earth

To understand the Cosmos it appears helpful to understand the origin of Organic Life on earth. Just like the Cosmos, Organic Life seems to have originated in a simple mix of chemicals on the early earth that combined to form primitive virus-like entities due to (perhaps) electrical charges such as lightning.

Under the stress of environment the primitive life evolved to become bacteria. After a time, further environmental stress caused bacteria to combine into small agglomerations of several bacteria that had enhanced survivability due to the interchange of nutrients.

Eventually the bacteria differentiated and through larger agglomerations became plants and animals. These entities further evolved to present day creatures.

This thumbnail description sets the stage for the consideration of the Cosmos. We have necessarily described the Cosmos sequentially starting from a point and progressing to spaces and then universes. The discussion led directly to our Unified Standard Model (UST), which encompasses what we know of Elementary Particles (often presented in the form of The Standard Model.) We were led to Cosmos theory by the development of the UST over a period of years. About six years ago it became clear to the author that the number of fundamental fermions in the UST suggested a basis in Cayley numbers. Over the subsequent years the Cayley number view blossomed into Cosmos Theory. (Cosmos Theory did not spring into existence magically – but rather through careful consideration.)

Having seen the progress of the Cosmos theory development we realize it has a logical sequential development as outlined in our recent book, *BEYOND TIME AND SPACE Fermion Strata & Consistency Conditions for Sequences, Cosmos Spaces and Gambols*.

The goal of this book is to present the development of Sequence Theory, Cosmos Theory and Gambol Theory in a logically based consequential form. In this respect it compares to Euclid's Geometry, which starts with definitions and five postulates and proceeds to develop Geometry logically but with a substantial introduction of additional concepts such as angles and figures as it progresses. The extension of Geometry is "natural" in the sense that the new features are simple extensions from the geometry generated from the postulates. For that reason one might call Euclidean Geometry a "Mathematical Life Form" since it starts from seeds (Postulates) and then grows extensions to form the mature subject.

We have followed the same general approach starting from a "seed" (a point and also totally antisymmetric tensors) and growing the Cosmos with its universes to a "mature" Cosmos extended to have many universe "branches." We have taken the same liberties, as Geometers, in developing our triad of theories and their universes. See Fig. 1.1.

1.1 Tracing Life Back From a Primordial Soup

We trace life back to a primordial soup. One might pursue the question of life much further back to Chemistry, atoms, fundamental particles, suns, galaxies, the universe and the Big Bang.

However that is not often done. One starts with the soup, and its constituents, and Chemistry because these aspects are complete enough to describe the Reality from which Life emerges. Exploring further back appears to be without value. (If alien life is found, then this view might require changes.)

We shall take the same view of Physics. We go back to a point whose completeness of description supports the development of a complete Physics of Reality. Before that, there is no compelling interest since it has no palpable consequences for the Physics of Reality. In saying that, there is a point to intellectual curiosity as long as it is realized that there are apparently no (significant) consequences for the Physics of Reality.

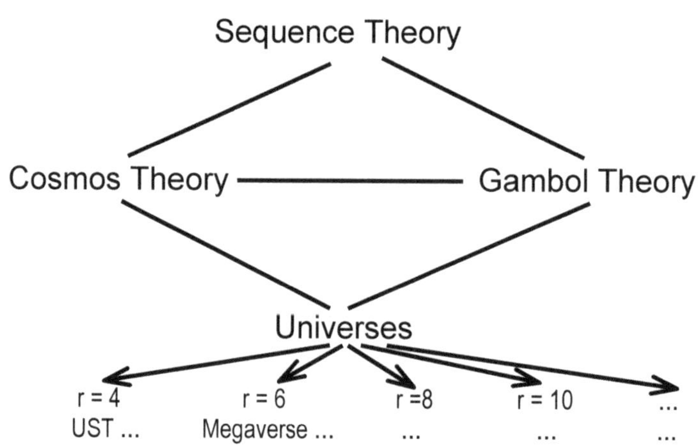

Figure 1.1. The Sequence, Cosmos and Gambol Theories combine to provide a complete, unified theory of fundamental Physical Reality. Their features interlock in many ways. Their approaches are compatible. Their consequences are decisive for the Cosmos.

2. The Purpose of Physics Theory

Since we are attempting an ultimate Physical Theory of Physical Reality it is appropriate to ask: What constitutes such a theory? How shall we describe it to meet our need to understand Physical Reality? Shall it be based on listing all possible equations of Physics? Or should it be based on a logical derivation from first principles after the manner of Euclidean Geometry?

2.1 What Constitutes Such a Theory?

The acid test of an ultimate theory is to answer all Physical Questions raised by Physical Experiments. An important caveat to this statement is the incalculability of answers to many theoretical questions due to the extreme difficulty of the calculations and/or the impossibility to calculate successfully on a computer with current computer technology. In these cases one must look to the envelope of results from a proposed ultimate theory to ascertain its overall success. Then acceptance of the theory becomes a gamble making it subject to further investigation, possibly refinement, and a continuing study of its rate of success – growing or not growing.

2.2 Our Need to Understand Physical Reality?

There are two purposes for the study of Physical Reality: to further the technical progress of Humanity helping it to overcome social and economic problems, *and* to advance the intellectual level of humanity raising it to a higher understanding of Reality.

These purposes are clearly being realized – perhaps more quickly than we anticipated.

2.3 A Basis of Physics by Listing all Possible Equations of Physics?

One way of obtaining the ultimate theory of Physics is to write down all possible equations governing Physical phenomena. Since this approach would imply listing all dynamic equations for the ultra-small to the ultra-large, and for the lowest energies to the highest energies, we clearly see that creating this *infinite* set of equations is an impossible task.

Thus Physics seeks to find the simplest most general theories, from which all equations can be derived as needed by technology and experiment. The goals of Simplicity and Generality are the premier hallmarks of a successful Physics.

2.4 A Basis for Physics in a Logical Derivation from First Principles

For good reason we seek to found a Physics on the basis of Simplicity and Generality. Physics is too big to be handled otherwise. Most Physicists have experienced the great growth of implications and computations that often followed from the simplest assumptions. Keeping the initial assumptions of a Physical system as simple and tight (least expensive in computation) is almost a universal goal of Physicists at work.

So we look to define a Physical System (such as the Cosmos) with a minimal structure and a minimal set of simple assumptions. We have developed Sequence Theory, Cosmos Theory and Gambol Theory based on that approach. We believe that we have succeeded – as the following chapters, and 12 years of our recent efforts show.

2.5 Physics is a Simulation

Physics and Reality are separate things. If one changes a Physics law or dynamics, whether it is a new innovation or a revision, there is no change in Reality.

There is no creation by pure thought.

To the extent that it is correct, a Physics[1] theory describes Physical Reality – the past, the present, and a projection into the future. The description provided by a Physics theory is meant to be a faithful reflection of Reality. For these reasons we can only view Physics as providing *simulations* of Reality. We must distinguish between the thing and its reflection. A simulation provides a picture – not the Reality.

Having recognized the nature of Physics we must remember its importance in describing natural phenomena and their dynamics. Physics has the great benefit of unifying our view of multi-faceted Reality.

2.6 Why is Physics Connected to Reality?

We know Physics is based on Mathematics. We know it can simulate Reality. Upon what is Reality based that it can be so directly simulated by Physics? The question is rather like a question of Cassius in *Julius Caesar*:[2] "Upon what meat does this our Caesar feed that he is grown so great?" Upon what meat does Reality feed that it is so greatly in tune with the Mathematics of Physics?

A rapprochement in understanding between Physics and Reality is still needed. The answers to these weighty questions are beyond Physics – it is for the future. To those who will answer it we can only say with Hamlet, "Thou, when thou return'st, wilt tell us." At the moment we must be satisfied with the answer: Reality is similarly Mathematical.

Could it have been otherwise? Reality might have been pure chaos.[3] But we see no evidence for it. Reality might have been a combination of a sector based on the application of Mathematical laws, and a chaos sector. Again we see no evidence for this possibility. We are left with the enthronement of a completely Mathematical Reality. Therefore Physics!

---

[1] Physics is a collection of theories and data in search of a unification into a complete simulation of Reality.

[2] W. Shakespeare, *Julius Caesar*, the play.

[3] By pure chaos we mean totally devoid of any Physical laws or structure. The best definition of chaos in this context is unregulated formlessness.

This result leads to the question: What form does the fundamental Mathematical formulation of Physics take? There are a number of possibilities. Our development of Sequence Theory and Cosmos Theory raises an interesting possibility. These theories lead directly to a structure based on powers of 2 and 4. Realizing that computer bits are also based on the integer 2, and computer programs similarly involve the manipulation of bits, we are led to consider the view that our theories ultimately follow from an underlying form of computation[4] – the source of Reality. This topic is currently being investigated by the author.

---

[4] Our book, Blaha (1998), *Cosmos and Consciousness* develops a relation between computer assembly language and quantum field manipulations. Some other related books are: Blaha (2005c), *The Equivalence of Elementary Particle Theories and Computer Languages: Quantum Computers, Turing Machines, Standard Model, Superstring Theory, and a Proof that Gödel's Theorem Implies Nature Must Be Quantum*; Blaha (2009), *The Algebra of Thought & Reality: The Mathematical Basis for Plato's Theory of Ideas, and Reality Extended to Include A Priori Observers and Space-Time Second Edition.*

3. A Beginning Where There is No Beginning

Most Physical Theories and studies have an initial starting point, (usually a time and Physical setup) and proceed to the consideration of the dynamical evolution of the Physical system.

However, our consideration of Physical Theories of the Cosmos is based on the absence of an initial time or final time. It is not a created entity. It is ever present.

3.1 Structure and Substance

We have suggested in previous studies that theories of the Cosmos benefit greatly from a separation into structure and substance. Cosmos Theory has a set of spaces. These spaces only specify arrays of dimensions. Nothing is within them. They are thus structural. They require no energy to create them.

3.2 Substance – Universes

On the other hand Cosmos Theory supports the idea of universes. Universes have substance. They have mass-energy when they are created. Cosmos Theory supports creation at specific times where time is measured within a Parent universe. The creation of the Parent universe (based on a Consistency Condition described later) gives a Parent time and set of coordinates. Later we describe the origin of the substance of universes based on a specific view of the nature of energy and mass.

3.3 Sequential Derivation or Ever Present Reality

Our development of Cosmos Theory and Sequence Theory is necessarily sequential – a logical derivation from initial postulates. However its reality is that it exists *ab initio. All features of their structure are present irrespective of any time.* The universes that occur appear in a Parent universe or sub-universe. Their appearance, and dynamical evolution, is determined by dynamical equations that most likely are in a Universe Quantum Field Theory. We have considered this possibility in earlier books.

3.4 Before the Conceptual Beginning

One might view the Cosmos as starting from a certain point in a certain configuration. This approach raises the question: What happened before that point?

We choose to specify Nothing! We do so later with an initial cosmos mathematical point that has *no vestige of prior* Physics.

We thus avoid an endless succession of pasts. Theories which start with a somewhat detailed preliminary setup must then account for the origin of the setup, and the origin of the origin, and so on to an infinite Physically unknown past. Our approach, presented later, is based on an absolute simplicity that requires no prior setup or knowledge.

3.5 Other Approaches to the Beginning

The answer to the question of a pre-Physics configuration often leads to a formulation that has no experimental consequences.

We suggest that features of a theory that have no experimental consequences can only be viewed as matters of "faith" – not in a religious sense but rather in a Philosophical sense. Thus it seems that other theories of the Cosmos with a prior beginning might be viewed as theoretically questionable. One cannot Physically leapfrog to a negative infinite time.

3.6 A Cosmos Theory With No Beginning

We have developed a theory of the Cosmos, which, for good reason, has no beginning time. The basis of this feature is my belief that a theory that "begins" with no coordinates, and particularly no time coordinate, cannot begin at a non-existent time. When coordinates are introduced through the appearance of a universe, then a "clock" may be started within the universe for the measure of time. (We assume that a Physical theory can only have a time coordinate(s) within a universe.)

3.7 Is There More?

In creating the Theories of Sequences and the Cosmos we have considered the question of a further more fundamental theory. We have been unable to explore further than these theories at present. The simplicity of their formulation, the Logic of their consequences, and the close predictions of known Elementary Particle experimental results argues for a finitude to further hypotheses. In saying that, we realize that there is no ultimate end to the growth of Physical understanding. We see a certain *infinity* in ultimate Physical knowledge that precludes a complete comprehension of the Physics of the Cosmos.

4. No End to the Cosmos

If there is no beginning, can there be an end?

We have seen Sequence Theory and Cosmos Theory are set without a location in time or space. They are perpetual. We have seen the structure of these theories which is conceptual and not requiring any setup or preconditions. Upon the entry of *substance*, which begins with universes, we encounter time and space within universes. Then space and time become of interest – but only within universes.

We thus see that Sequence Theory and Cosmos Theory structure are without an end while universes may end with one exception – the Parent universe. There may be more than one Parent universe(s). We assume there is only one Parent universe. If there were more Parent universes we would treat them analogous.

The Parent universe results from a Consistency Condition that we have developed previously. Its essence is the virtual forces that exist due to an outward thermodynamic pressure and an inward directed Casimir force due to a virtual vacuum. The outward pressure would cause expansion. The inward Casimir pressure would cause contraction. The Parent universe is the least dimension universe that is stable or expands. A lower dimension universe would contract to nothingness and thus be non-existent. An expanding (or stable) universe continues to exist.

How long can it exist? There is no fixed lifetime. So it cannot end! The sub-universes within it are created and possibly destroyed according to their dynamics. In previous books we raised the possibility of a Quantum Theory of universes with interactions between them and a capability for collisions and amalgamation.

We conclude structure is forever; the Parent universe exists forever, sub-universes are created and destroyed according to universe dynamics. They may exist forever in Parent universe time or may have a finite lifetime.

From a Point to the Cosmos Spaces

5. Primordial Point for Structure

There are many points.

As we saw earlier we differentiate between our description of Sequence Theory and Cosmos Theory structure which is sequential and logical, and the Physical Reality of their features which has an *ever-present* conceptual existence – they are independent of any time in the absence of time.

We attempt to begin the *sequential description* of the structure of these theories in the simplest manner. We specify their "beginning" with a mathematical point with dimension zero. This point is devoid of all features. It has no predecessor. Thus there is no doorway to a past. As we stated earlier there is no past since there is no time. We avoid a recurring attempt to find a past.

In a manner analogous to the development of Geometry we next develop a construction process for the structure of Sequence Theory and Cosmos Theory. We use the potentialities inherent in the concept of dimension to generate the construction process. Later in chapter 22 we will see these features are embodied in the Master Equation.

5.1 Dimension

The concept of dimension is best viewed as a primitive term. (Geometry has primitive terms as well. Non-Euclidean Geometry is a consequence of a different choice of primitives.)

We assume a dimension is an item (a primitive) that has no numeric value and no other type of value. We assume dimensions are countable although the possibility of an infinite dimension is allowed. We may group dimensions into sets.

With that basis we are now able to engage in a construction process leading to the dimensions of spaces.

6. Growth to a Two Dimension Space

This chapter[5] grows a Mathematical point to a two dimension space of infinite extent. It begins by growing the point to a straight one dimension line segment. Then a construction of a two dimension unit grid follows along the lines of the Hilbert fractal curve construction. Our construction generates the Cosmos Theory fractal Curve.

The generated unit square is then extended to an infinite two dimension space.

6.1 Generation of Line Segment

The Cosmos Curve is constructed from a mathematical point to a unit line segment and then to a two-dimension unit square grid. The construction from a mathematical point to a line segment relies on the sum:

$$\sum_{n=1}^{\infty} 2^{-n} = 1 \tag{6.1}$$

The terms in this summation correspond directly to ½ of the length of corresponding columns d_c in the Cosmos Theory space spectrum of Limos. One views the generated line segment as the combination of individual line segments of lengths ½, ¼ , … $1/\infty$ giving a unit length line. The initial point is the $1/\infty$ term. The sum of the line elements of length 2^{-n} is a line of unit length.

6.2 Generation of Fractal Curve Giving a Unit Square

We now construct the remainder of the Fractal Cosmos Curve starting from the unit line segment by combining instances of multiples of this curve of length 2^n to generate a space filling unit square grid – a fractal curve. This square has two dimensions. It may be characterized by x and y coordinates. The Fractal Cosmos Curve is a mathematical construct. We "smooth" the curve to a two dimension unit square.

6.3 Two Dimension Unit Square

We define a coordinate system for the unit square with coordinates x and y:

$$\begin{aligned} &0 \le x \le 1 \\ &0 \le y \le 1 \\ &0 \le x + y \le 2 \end{aligned} \tag{6.2}$$

6.4 Extension of Unit Square to an Infinite Space

Having defined the unit square we now simply extend the limits of each coordinate from $-\infty$ to ∞. Thus we now have a construction process from a zero dimension point to a two dimension space of infinite extent. In chapter 7 we will define

---

[55] Previous books present more detail on the contents of this chapter. See for example the Appendix of Blaha (2025c).

a Master Equation on that space that leads to Sequence Theory and then to Cosmos Theory and its set of spaces.

6.5 A Conceptual Derivation

While the above derivation is Mathematically correct it is important to realize it does not create a material entity. It "creates" an abstraction – a structure for the Cosmos Theory. No energies are involved. No substance is created. It is Form without material substances.

7. The Master Equation of Physical Reality

Having obtained a two dimension space (with no time variable) we now turn to developing an equation that describes a spectrum of states. We viewed this space in chapter 6 as originating in a *mathematical point*. We obtained this infinite space through scaling up the unit square. We call this space the Fractal Cosmic Curve space.[6] (The Cosmic Curve is similar in construction to the Hilbert fractal curve.)

The scaling feature is critical to the creation of the space. We therefore require it in the Master Equation that we develop to generate the Cosmos set of spaces. We specify it with a simple Schrödinger-like equation in coordinates x and y with a λ/r^2 scaling potential. The scaling potential provides the simplest guarantee of an energy spectrum that is a sequence of powers of a constant.[7]

$$[(\partial^2/\partial x^2 + \partial^2/\partial y^2)/m + \lambda/m^2r^2 + C)\psi = 0 \qquad (7.1)$$

where λ is a coupling constant and $r = (x^2 + y^2)^{1/2}$. Using polar coordinates we have

$$x = r\cos(\varphi)$$
$$y = r\sin(\varphi)$$

and find

$$(\partial^2/\partial r^2 + 1/r\ \partial/\partial r + \partial^2/\partial\varphi^2 + \lambda/mr^2 + mC)\psi = 0 \qquad (7.1a)$$

\

$$(\partial^2/\partial\varphi^2)\psi = 0$$

with the solution

$$\psi = \psi(r)\Phi(\varphi) \qquad (7.1b)$$

where

$$\Phi(\varphi) = \exp(ik\varphi) \qquad (7.1c)$$

The quantum number k must be a whole integer to avoid irregularities in $\psi(r)$.

Starting from eq. 7.1a which implies

$$(\partial^2/\partial r^2 + 1/r\ \partial/\partial r + (\lambda - k^2)/mr^2 + mC)\psi = 0 \qquad (7.1d)$$

we eliminate the $\partial/\partial r$ term by letting

$$\psi(r) = r^{-1/2}\ w(r) \qquad (7.1e)$$

to obtain

$$w'' + ((\lambda' - k^2)/r^2 + mC)w = 0 \qquad (7.1f)$$

---

[6] This is detailed in Blaha (2024c), *Fractal Cosmic Curve: Tensor-Based Cosmos Theory.*

[7] This equation may be viewed within the context of conformal symmetry due to its local scale invariant potential.

where

$$\lambda' = \lambda + 0.25$$

We now let

$$\lambda = k^2 - 0.25 + n_\varphi{}^2$$

with the result

$$(\partial^2/\partial r^2 + \lambda_{n_\varphi}/mr^2 + mC)\, w(r) = 0 \qquad (7.1g)$$

where

$$\lambda_{n_\varphi} = n_\varphi{}^2/m \qquad (7.1h)$$

The eigenvalue spectrum due to eq. 7.1h is[8]

$$C_n = \text{const}\, \exp(-n\pi/(\lambda_{n_\varphi} - \tfrac{1}{4})^{\frac{1}{2}}) \qquad (7.2)$$

where n and n_φ are independent quantum numbers.[9] The quantization of the C_n eigenvalue spectrum is based on a requirement of orthogonality of solutions (an observation due to Von Neumann).

Thus we initially have a radial energy series in eq. 7.2. We generalize this series to a double series in n and an angular momentum quantum number n_φ.

$$C_{n_\varphi n} = c_{n_\varphi} \exp(n\pi/(\lambda_{n_\varphi} - \tfrac{1}{4})^{\frac{1}{2}}) \qquad (7.2a)$$

where n is the radial quantum number and n_φ is the angular momentum quantum number. The growth factor for each power series sequence is

$$g_{n_\varphi} = \exp(\pi/(\lambda_{n_\varphi} - \tfrac{1}{4})^{\frac{1}{2}}) \qquad (7.2b)$$

The radial quantum number, n, parameterizes each sequence of "energy" values. The angular momentum quantum number n_φ, which we define in the next section, specifies the sequence and its constant factor c_{n_φ}. The index n_φ determines the coupling constant $\lambda - \lambda_{n_\varphi}$ in eq. 7.1, for each specific sequence.

We will see each space in the set of 10 HyperCosmos spaces (Fig. 7.1) defines the size of the dimension array size d_{dr} of of the space:

$$d_{dr} = 2^{r+4} \qquad (7.2c)$$

using $g_{n_\varphi} = 2$ for $n_{n_\varphi} = 4$. Introducing the concept of strata for fermions seen later, where each fermion becomes one of four strata fermions, increases the dimension size to

$$d_{\text{strata dr}} = 2^{r+6} \qquad (7.2d)$$

[8] The change to w(r) leads to inconsequential differences with the calculation in our previous books. The new form of the differential equation is like a quantum mechanical Schrödinger radial equation.
[9] Based on K. M. Case, Phys. Rev. **80**, 797 (1950).

We consider the cases of other sequences (Fig, 7.2) in the following chapters.

THE HYPERCOSMOS SPACES SPECTRUM

Blaha Space Number $N = o_s$	Cayley-Dickson Number n	Cayley Number Cn d_c	Dimension Array column length $d_{cd}=d_{cr}$	Dimension Array Size $d_{dN}=d_{dr}$	Space-time-Dimension r	CASe Group $su(2^{r/2},2^{r/2})$ CASe
0	10	1024	2048	2048^2	18	su(512,512)
1	9	512	1024	1024^2	16	su(256,256)
2	8	256	512	512^2	14	su(128,128)
3	7	128	256	256^2	12	su(64,64)
4	6	64	128	128^2	10	su(32,32)
5	5	32	64	64^2	8	su(16,16)
6	4	16	32	32^2	6	su(8,8)
7	**3**	**8**	**16**	$\mathbf{16^2}$	**4**	**su(4,4)**
8	2	4	8	8^2	2	su(2,2)
9	1	2	4	4^2	0	su(1,1)

Figure 7.1. The Cosmos Theory HyperCosmos space spectrum. Note: $n_\varphi = 4$. The parameter n is the Cayley-Dickson number. See Blaha (2022c).

	Sequence	n_φ	g_g multiplier factor	λ_{n_φ}
A	Cosmos dimension array column lengths $2^{r/2+2}$	4	2	20.79
B	Coupling Constants Multiplier	2	4	5.39
C	Fermion Sequence 2 Masses	1	$2/3 \times 2^5$	1.3
D	Fermion Sequence 1 Masses	$½^{½}$	$3/2 \times \pi 2^5$	0.65
E	n_φ Values Sequence; A Sequence of Sequences	½	95939	0.325

Figure 7.2. Cosmos Theory Sequences. Note: The fermion Mass sequences must apply to mass sequences in other generations and other layers up to overall factors. The coupling constant sequence must apply to the coupling constant sequences in other layers up to overall factors. The values of n_φ, g_g, and λ_{n_φ} should apply for all generations and layers. The sole difference is the values of the starting sequence values $c_{n_\varphi n}$ from eq. 7.2a. Note $\boldsymbol{\lambda_{n_\varphi}} > 0.25$ is required by eq. 7.1 in order to have the bound state energy spectrum. Thus the possible additional line with $\mathbf{n_\varphi} = ¼$ is ruled out.

8. Cosmos Spaces

The discussion of the structure and spectrum of Cosmos spaces begins with the sequence of spaces indicated by row A of Fig. 7.2. We see spaces are a series of entities with arrays of dimensions that have dimensions characterized by a multiplier, 2. These arrays we will take to be arrays of dimensions. In making this interpretation we are guided by a parallel derivation based on 1. The numerics of totally antisymmetric tensors are guided by a close analogy with Dirac γ matrices and 2. The numerics of creation/annihilation operators in Fourier expansions of PseudoQuantum fermion wave functions.[10]

8.1 Derivation of HyperCosmos Spaces' Dimension Arrays

The derivation of Cosmos space dimension arrays, which are the sole content of a Cosmos space, begins with the derivation of Dirac γ matrices. Dirac γ matrices are square arrays. Their column length and row length follow from the number of independent totally antisymmetric tensors of dimension r:

$$\text{Number of tensors} = \text{column length} = 2^{r/2} \tag{8.1}$$

The number of components of a Dirac γ matrix is then

$$2^r \tag{8.2}$$

We now require a Cosmos dimension array is similar to a Dirac γ matrix up to a constant. We determine the size of a Cosmos dimension array column by requiring that it has the same size as the number of creation and annihilation operators in Fourier expansions of PseudoQuantum fermion wave functions.[11] This requirement sets the size (number of components) in the dimension array to

$$d_{dr} = 2^{r+4} \tag{8.3}$$

as shown in Fig. 7.1. Since the number of components in a Dirac γ matrix is the same for dimension r and dimension r + 1, we limit the set of values of r to be even numbered – spaced by units of 2. The result is the table in Fig. 7.1. Thus we have determined the spectrum of HyperCosmos spaces and their dimension arrays. *We view dimension arrays as an Internal Symmetry equivalent to Spin-related Dirac γ matrices.*

[1010] The figures and text in this section are based on Blaha (2023d), *Newton's Apple is Now the Fermion,* and other books by the author. These books provide additional detail and discuss additional topics.

[11] See chapter 4 of Blaha (2023c) and earlier books for more details.

8.1.1 The Explicit Determination of the Size of the Dimension r Dimension Array

The size of the dimension array can be determined from the number of creation and annihilation operators in the r dimension PseudoQuantum fermion field operator.

First we note that in a space-time of dimension r there are $2^{r/2-1}$ spin values for each b type operator. Then we multiply by a factor of 2 to include the d type operators. Since the operators are complex-valued we multiply by 2 to bring in the conjugate operators $b^\dagger$ and $d^\dagger$ operators. The result is $2^{r/2+1}$ b and d type operators in a standard Fourier representation of a fermion field operator.

Lastly, we specify PseudoQuantum field operators, which brings in another factor of 2. The result is the number of b's and d's for each momentum taking account of all spins and the PseudoQuantum fields is $2^{r/2+2}$. This value is the dimension array column length.

The number of dimension array components is the square: eq. 8.3 above and in previous books by the author. The d_{dr} array sizes appear in Fig. 7.1.

8.1.2 The r = 4 Space of Our Universe

The dimension array for our universe is of r = 4 dimensions. It is a square array of 16×16 rows and columns. The array is mapped to a set of four layers of fundamental fermions with each layer consisting of four generations of fermions. Each generation has eight Normal fermions and eight Dark fermions. Fig. 8.1 shows the spectrum of fermions.

The array can also be mapped to Symmetry Groups. There is a set of symmetry groups for each layer as shown in Fig. 8.2. Note the SL(2, C) group for the four coordinates of the universe. Note also the appearance of Generation Groups and Layer Groups. In each layer the Generation groups mix the four generations of fermions (separately for Normal and Dark fermions.) The Layer groups are placed in layers although they mix fermions in different layers. Fig. 8.3 displays Layer group interactions. See chapters 7 and 8 of Blaha (2024ℓ) for a detailed discussion of these groups and their matrix representations.

Number of Columns =	4 NORMAL	4	4 DARK	4
Layer 1 4 generations	**e 3 up-quarks**	**ν 3 down-quarks**	**e 3 up-quarks**	**ν 3 down-quarks**
Layer 2 4 generations	**e 3 up-quarks**	**ν 3 down-quarks**	**e 3 up-quarks**	**ν 3 down-quarks**
Layer 3 4 generations	**e 3 up-quarks**	**ν 3 down-quarks**	**e 3 up-quarks**	**ν 3 down-quarks**
Layer 4 4 generations	**e 3 up-quarks**	**ν 3 down-quarks**	**e 3 up-quarks**	**ν 3 down-quarks**

Figure 8.1. The r = 4 16×16 fermion spectrum of our universe tentatively arranged as SU(4)-plets that correspond directly with SU(4) (or SU(3)⊗U(1)) fermions. There are four layers. Each set of 4 fermions has 4 generations matching the number of rows in each layer. This Periodic Table is broken into Normal and Dark sectors. The electron type fermions are denoted **e**. The neutrino type fermions are denoted **v**. Each layer has four generations. **The generations depicted here have a different form compared to those of The Standard Model.** This view of the fermion spectrum changes when strata are introduced later.

NORMAL

SU(3)⊗U(1)	SU(2)⊗U(1)⊗SL(2, C)
Generation U(4)	Layer U(4)
SU(3)⊗U(1) Generation U(4)	SU(2)⊗U(1)⊗U(2) Layer U(4)
SU(3)⊗U(1) Generation U(4)	SU(2)⊗U(1)⊗U(2) Layer U(4)
SU(3)⊗U(1) Generation U(4)	SU(2)⊗U(1)⊗U(2) Layer U(4)

DARK

SU(3)⊗U(1) Generation U(4)	SU(2)⊗U(1)⊗U(2) Layer U(4)
SU(3)⊗U(1) Generation U(4)	SU(2)⊗U(1)⊗U(2) Layer U(4)
SU(3)⊗U(1) Generation U(4)	SU(2)⊗U(1)⊗U(2) Layer U(4)
SU(3)⊗U(1) Generation U(4)	SU(2)⊗U(1)⊗U(2) Layer U(4)

Figure 8.2. The r = 4 symmetry groups (UST). The Dark parts have not as yet been found. Note each element has a 16 real dimension representation. The SL(2, **C**) representation has four coordinates.[12]

[12] The Lorentz Group $SO^+(1, 3)$ is often specified with an SL(2, **C**) representation.

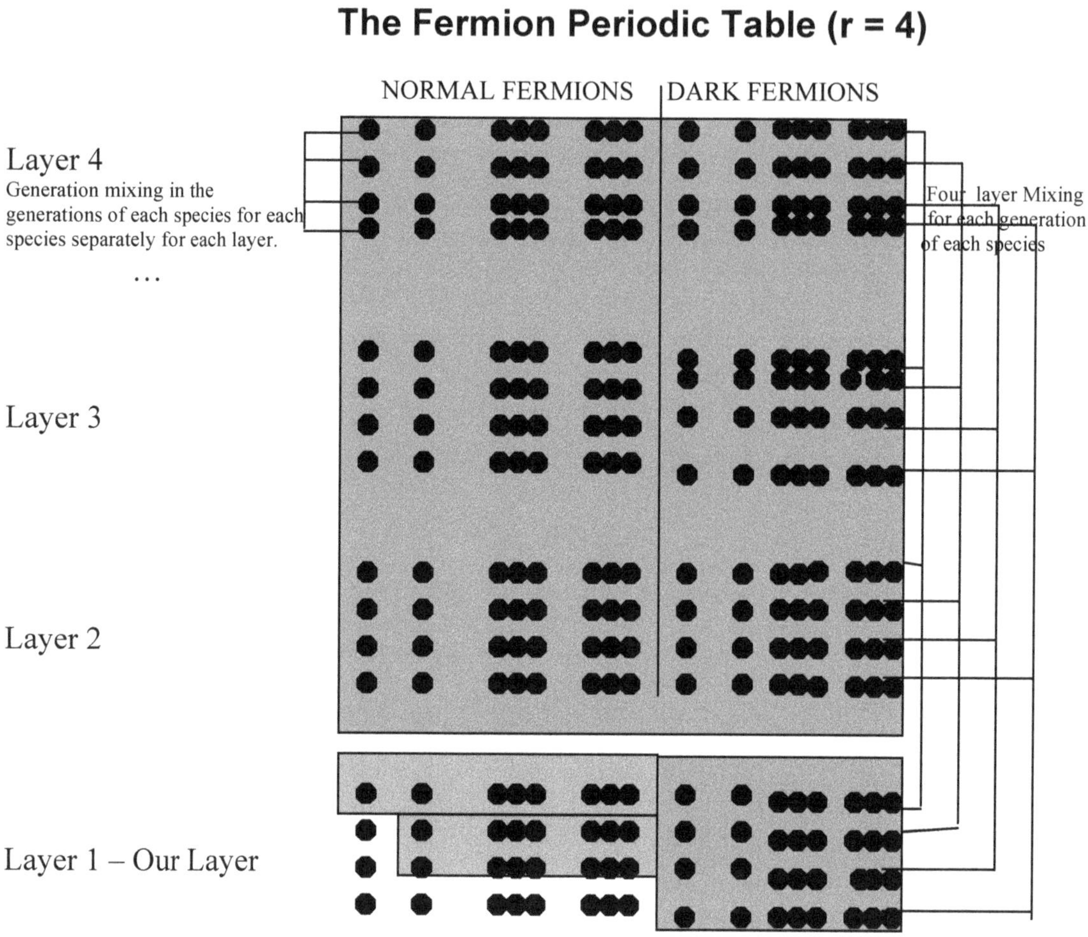

Figure 8.3. The r = 4 Fermion particle spectrum and partial examples of the pattern of mass mixing of the Generation groups and of the Layer groups. Unshaded parts are the known fermions, One shaded generation is an additional, as yet not found, 4th generation. The lines on the left side (only shown for one layer) display the Generation Group mixing within each layer. The Generation mixing occurs within each layer using a separate Generation group for each layer. The lines on the right side show Layer group mixing (for Dark matter) with the mixing among all four layers for each of the four generations individually. There are four Layer groups for Normal matter and four Layer groups for Dark matter. There are 256 fundamental fermions. This gives some UST interactions and the UST fermion spectrum. This view of the fermion spectrum changes when strata are introduced later.

8.2 HyperCosmos Spaces of the Second Kind

The HyperCosmos spectrum of spaces presented above has both a Normal and a Dark sector. Since the existence of the Dark sector is a matter of controversy at present, we now give a variation on the spaces spectrum that omits the Dark sector. This changed spectrum is obtained by omitting the d's from the calculation in section 8.1.1.[13] The result is the HyperCosmos of the Second Kind with the dimension array size:

[13] It is somewhat remarkable that the omission of the "negative energy" fermion operators is tied to the omission of the fermion and symmetry group Dark sector.

$$d_d = 2^{r+3} \tag{8.4}$$

as in Fig. 8.4.

The absence of the Dark sector has significant results. They are evident in the Second Kind HyperCosmos spaces of Fig. 8.5 – 8.8.

One significant result is: there are no "horizontal" U(2) Connection groups labeled A, B, C, and D in Fig. 8.8. The U(2) "vertical" Connection groups E, F and G are replaced by SL(2, **C**) coordinate space groups in our quadplex formalism with y, z, u, and v coordinate systems.

HYPERCOSMOS OF THE SECOND KIND SPACES SPECTRUM

Blaha Space Number $N = O_s$	Cayley-Dickson Number n	Cayley Number d_c	Dimension Array size d_d	Space-time-Dimension r	CASe Group $su(2^{r/2}, 2^{r/2})$ CASe
0	10	1024	1024 × 2048	18	su(512,512)
1	9	512	512 × 1024	16	su(256,256)
2	8	256	256 × 512	14	su(128,128)
3	7	128	128 × 256	12	su(64,64)
4	6	64	64 × 128	10	su(32,32)
5	5	32	32 × 64	8	su(16,16)
6	4	16	16 × 32	6	su(8,8)
7	**3**	**8**	**8 × 16**	**4**	**su(4,4)**
8	**2**	4	4 × 8	2	su(2,2)
9	**1**	2	2 × 4	0	su(1,1)

Figure 8.4. The HyperCosmos of the Second Kind space spectrum. The r = 4 space for our universe is in bold type. Note the changed d_d column relative to the HyperCosmos. Note the array sizes. The arrays are not square.

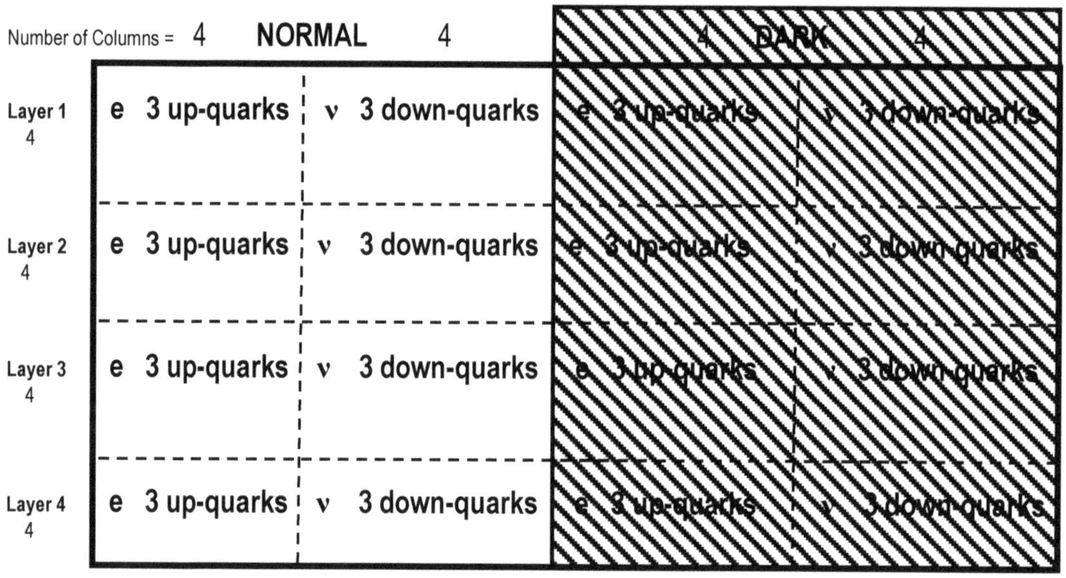

Figure 8.5. The HyperCosmos of the Second Kind 8×16 fermion spectrum tentatively arranged as SU(4)-plets that correspond directly with SU(4) (or SU(3)⊗U(1)) fermions. There are four layers. Each set of 4 fermions has 4 generations matching the number of rows in each layer. This Periodic Table is broken into Normal and Dark sectors. The absence of a Dark sector is indicated by the darkening. This view of the fermion spectrum changes when strata are introduced later.

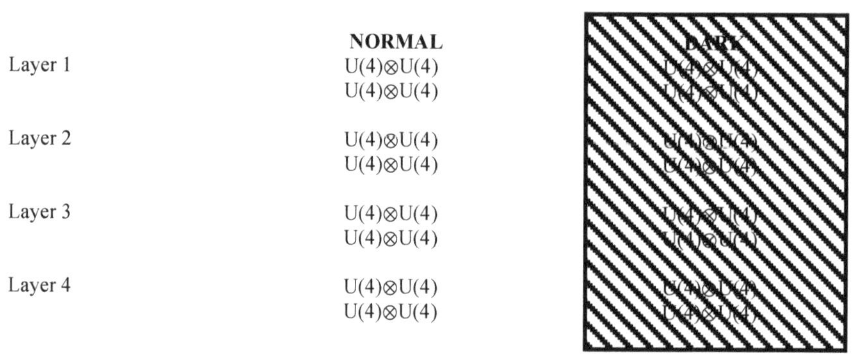

Figure 8.6. The r = 4 Second Kind "initial" distribution of sets of symmetry groups. Each set is distinct and supports interactions only for the corresponding set of fermions (separately for Normal and Dark fermions). *Thus each set of 16 fermion generations has its own quantum numbers and interactions.* Each U(4)⊗U(4) set has a 16 real-valued dimension representation, which are of importance when we consider Fundamental Reference Frames. The absence of a Dark sector as indicated by the darkened part.

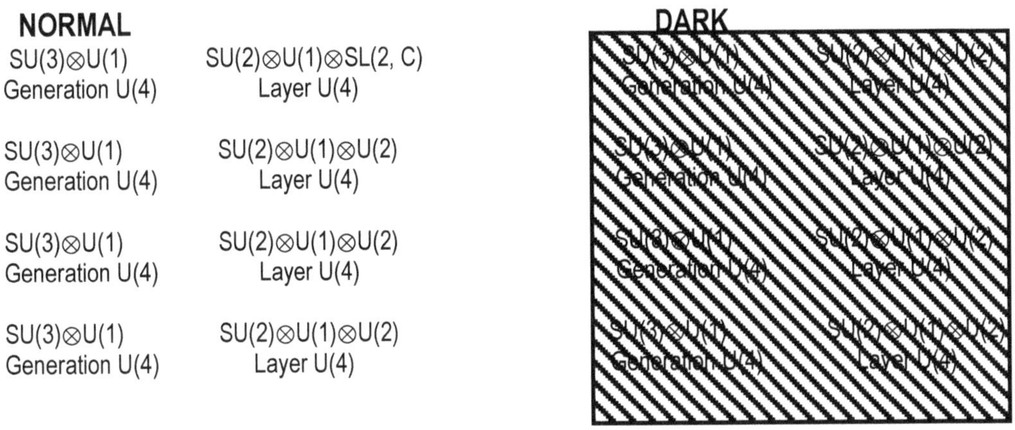

Figure 8.7. The transformed/broken sets of symmetries in the r = 4 dimension Second Kind HyperCosmos space. Note each element has a 16 real dimension representation. This depiction is also evident in a Second Kind UST. The SL(2, **C**) representation has four coordinates.[14] The absence of a Dark sector as indicated by the darkened part.

---

[14] The Lorentz Group $SO^+(1, 3)$ is often specified with an SL(2, **C**) representation.

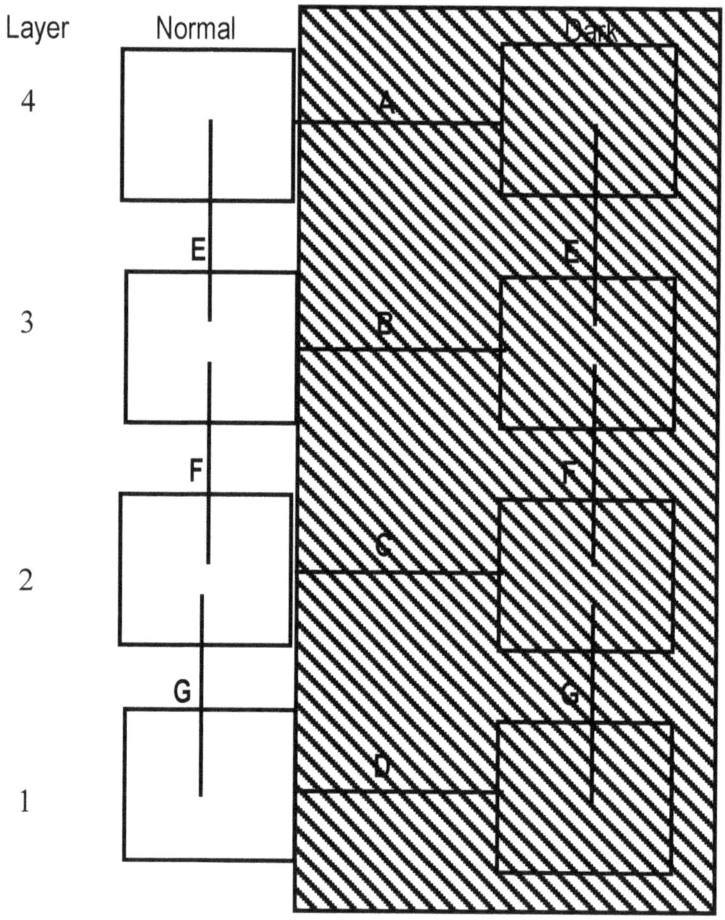

Figure 8.8. The three U(2) Connection groups[15] (shown as 3 lines) between the eight UST blocks in the r = 4 dimension Second Kind HyperCosmos. The Darkened part is not present in the Second Kind case. Connection groups are obtained by transfering 12 dimensions from the UST space-time to internal symmetries with the consequent reduction of the space-time from four octonion (complex quaternion) coordinates to four real coordinates. The Connection groups generate rotations and interactions between corresponding fermions and vector bosons of each pair of blocks.

8.3 Limos Spaces

The Limos spaces of Cosmos Theory are spaces with negative dimension r that are not physical spaces in the usual sense. See Fig. 8.9. We have described them as describing constituents of fundamental fermions. We have called these constituents gambols and successfully applied the concept to the consideration of particle decays and interactions.[16]

---

[15] Connection groups are discussed in Appendix 9-B.

[16] See Blaha (2023e), *Cosmos Theory: The Sub-Particle Gambol Model*; and Blaha.(2024a), *Cosmos-Universe-Particle-Gambol Theory*.

Later we will reconsider the interpretation of gambols in terms of fermion strata. See section 10.2 on Gambol Strata.

THE Limos SECTOR SPACES SPECTRUM

10	0	1	2	2^2	-2	U(1)	-3/8	0	0
11	-1	½	1	1^2	-4	U(½)	-7/16	-4	0
12	-2	¼	½	$½^2$	-6	U(¼)	-15/32	-8	0
13	-3	1/8	¼	$¼^2$	-8	U(1/8)	-31/64	-12	0
14	-4	1/16	1/8	$1/8^2$	-10	U(1/16)	-63/128	-16	0
				•					
				•					
				•					
∞	−∞	0	0	0	−∞	U(0)	0	−∞	0

Figure 8.9. The Cosmos Theory space spectrum for negative dimension r. Spaces with non-negative space-times may have instances. Spaces with negative space-times form the Limos set of Limos spaces.

8.4 Unification Spaces

There are a number of forms of unification. We chose to define unification for Cosmos Theory as the capability to rotate internal symmetries in each space. We will show that a mechanism exists supporting this form of unification based on combinations of Internal Symmetry and General Relativistic transformations.

8.4.1 The Map from a Space to its HyperCosmos HyperUnification Space

Consider an r dimension space. It has a square dimension array that has $2^{r/2+2}$ columns and $2^{r/2+2}$ rows with a total of 2^{r+4} components (omitting strata). This array specifies the fermions of universes of the space (Fig. 8.1). It also specifies the total of the Internal Symmetry irreducible representation dimensions (Fig. 8.2).

We wish to specify a set of rotations[17] that rotate the fermions and Internal Symmetry dimensions. Some time ago this author showed how a General Relativistic transformation "rotates" the creation and annihilation operators of a quantum field. See the peer-reviewed papers:

S. Blaha, Phys. Rev. **D17**, 994 (1978).
S. Blaha, "The Local Definition of Asymptotic Particle States", IL Nuovo Cimento **49A**, 35 (1979)
S. Blaha, "New Framework for Gauge Field Theories", IL Nuovo Cimento **49A**, 113 (1979).

The rotations for scalar and fermion fields may be summarized with:[18]

The *most* general mode expansion of the fields in the coordinate system A with Fourier momentum α is

$$\varphi_{1A}(x) = \Sigma_\alpha\, [a_{1\alpha}\, f_\alpha(x) + a^\dagger{}_{1\alpha}\, f_\alpha{}^*(x)] \qquad (4.11)$$
$$\varphi_{2A}(x) = \Sigma_\alpha\, [a_{2\alpha} f_\alpha(x) + a^\dagger{}_{2\alpha}\, f_\alpha{}^*(x)] \qquad (4.12)$$

[17] We use rotation and transformation interchangeably. The above papers show, under General Relativistic transformations, fermion creation/annihilation operators transform in a manner analogous to rotations.
[18] From Blaha (2022e). Note equation numbers are those of that book.

and in coordinate system B with Fourier momentum β is

$$\varphi_{1B}(x) = \Sigma_\beta\, [a_{1\beta}\, f_\beta(x) + a^\dagger_{1\beta}\, f_\beta{}^*(x)] \qquad (4.13)$$
$$\varphi_{2B}(x) = \Sigma_\beta\, [a_{2\beta}\, f_\beta(x) + a^\dagger_{2\beta}\, f_\beta{}^*(x)] \qquad (4.14)$$

The creation/annihilation operators of A have the commutation relations:

$$[a_{i\alpha'}, a_\alpha] = 0 \qquad (4.15)$$
$$[a^\dagger_{i\alpha'}, a^\dagger_{j\alpha}] = 0$$
$$[a_{i\alpha}, a^\dagger_{j\alpha'}] = (1 - \delta_{ij})\delta^3(\boldsymbol{\alpha} - \alpha') \qquad (4.16)$$

for i, j = 1, 2. Two related vacuums are defined: $|0\rangle_1$ and $|0\rangle_2$ by

$$a_{1\alpha}|0\rangle_{2\alpha} = a^\dagger_{1\alpha}|0\rangle_{2\alpha} = 0 \qquad (4.17)$$
$$a_{2\alpha}|0\rangle_{1\alpha} = a^\dagger_{2\alpha}|0\rangle_{1\alpha} = 0$$
$$a_{i\alpha}|0\rangle_{i\alpha} \neq 0 \qquad a^\dagger_{i\alpha}|0\rangle_{i\alpha} \neq 0 \qquad (4.18)$$

for i = 1, 2 and all α and α′. And similarly for the creation/annihilation operators of B with α replaced with β.

The general coordinate system B Fourier expansion in terms of A creation/annihilation operators is

$$\varphi_{1B}(x) = \Sigma_\beta\, \Sigma_\alpha\, [(c_{11}a_{1\alpha} + c_{12}a_{2\alpha} + C_{11}a^\dagger_{1\alpha} + C_{12}a^\dagger_{2\alpha})f_\beta(x) + + (c'_{11}a^\dagger_{1\alpha} + c'_{12}a^\dagger_{2\alpha} + C'_{11}a_{1\alpha} + C'_{12}a_{2\alpha})f_\beta{}^*(x)] \qquad (4.19)$$

$$\varphi_{2B}(x) = \Sigma_\beta\, \Sigma_\alpha\, [(c_{21}a_{1\alpha} + c_{22}a_{2\alpha} + C_{21}a^\dagger_{1\alpha} + C_{22}a^\dagger_{2\alpha})\, f_\beta(x) + + (c'_{21}a^\dagger_{1\alpha} + c'_{22}a^\dagger_{2\alpha} + C'_{21}a_{1\alpha} + C'_{22}a_{2\alpha})f_\beta{}^*(x)] \qquad (4.20)$$

where the c_{ij} and C_{ij}, and c'_{ij} and C'_{ij} are all functions of α.

We define sets of Bogoliubov transformations B_1 and B_2 that give

$$a_{i\beta} = \Sigma_\alpha\, (f_\beta, f_\alpha)B_{2\alpha}B_{1\alpha}a_{i\alpha}(B_{1\alpha}B_{2\alpha})^{-1} = \Sigma_\alpha\, (f_\beta, f_\alpha)[c_{i1}a_{1\alpha} + c_{i2}a_{2\alpha} + C_{i1}a^\dagger_{1\alpha} + C_{i2}a^\dagger_{2\alpha}] \qquad (4.21)$$

$$a^\dagger_{i\beta} = \Sigma_\alpha\, (f^*_\beta, f^*_\alpha)\, B_{2\alpha}B_{1\alpha}a^\dagger_{i\alpha}(B_{1\alpha}B_{2\alpha})^{-1} = \Sigma_\alpha\, (f^*_\beta, f^*_\alpha)[c'_{i1}a^\dagger_{1\alpha} + c'_{i2}a^\dagger_{2\alpha} + C'_{i1}a_{1\alpha} + C'_{i2}a_{2\alpha}] \qquad (4.22)$$

8.4.1.1 The B_1 Transformation

The B_1 transformation has the form:

$$B_{1\alpha}(x_1, x_2) = \exp[x_1\Gamma^1{}_{3\alpha}]\, \exp[x_2\Gamma^1{}_{2\,\alpha}] \qquad (4.23)$$

where the Hermitian operators $\Gamma^1{}_{1i\alpha}$ are

$$\Gamma^1{}_{3\alpha} = (a^\dagger_{2\alpha}a_{1\alpha} + a_{2\alpha}a^\dagger_{1\alpha})/2 \qquad (4.24)$$
$$\Gamma^1{}_{2\alpha} = i(a^\dagger_{2\alpha}a^\dagger_{1\alpha} - a_{2\alpha}a_{1\alpha})/2 \qquad (4.25)$$

We also define

$$\Gamma^1{}_{1\alpha} = -\,(a^\dagger_{2\alpha}a^\dagger_{1\alpha} + a_{2\alpha}a_{1\alpha})/2 \qquad (4.26)$$

and

$$\Gamma^1{}_{4\alpha} = i(a^\dagger_{2\alpha}a_{1\alpha} - a_{2\alpha}a^\dagger_{1\alpha})/2 \qquad (4.27)$$

Together the three operators satisfy su(1,1) algebra commutation relations:

$$[\Gamma^1{}_{1\alpha}, \Gamma^1{}_{2\alpha'}] = -i\delta_{\alpha\alpha'}\,\Gamma^1{}_{3\alpha} \qquad [\Gamma^1{}_{2\alpha}, \Gamma^1{}_{3\alpha'}] = i\delta_{\alpha\alpha'}\,\Gamma^1{}_{1\alpha} \qquad [\Gamma^1{}_{3\alpha}, \Gamma^1{}_{1\alpha'}] = i\delta_{\alpha\alpha'}\,\Gamma^1{}_{2\alpha} \qquad (4.28)$$

They transform $a_{i\alpha}$ and $a^\dagger_{i\alpha}$ to terms of the form $c_{i\alpha}a_{i\alpha} + c_{i\alpha}{}'a^\dagger_{i\alpha}$ where $c_{i\alpha}$ and $c_{i\alpha}{}'$ are constants.

Some properties of the su(1,1) algebra are detailed in S. Blaha, "The Local Definition of Asymptotic Particle States", IL Nuovo Cimento **49A**, 35 (1979).

8.4.1.2 The B_2 Transformation

The B_2 transformation has the form:

$$B_{2\alpha}(x_1, x_2) = \exp[x_1\Gamma^2{}_{3\alpha}]\exp[x_2\Gamma^2{}_{2\alpha}] \quad (4.29)$$

where the Hermitian operators $\Gamma^2{}_{1i\alpha}$ are

$$\Gamma^2{}_{3\alpha} = (a^\dagger{}_{1\alpha}a_{1\alpha} + a_{1\alpha}a^\dagger{}_{1\alpha})/2 \quad (4.30)$$
$$\Gamma^2{}_{2\alpha} = (a^\dagger{}_{2\alpha}a_{2\alpha} + a_{2\alpha}a^\dagger{}_{2\alpha})/2 \quad (4.30)$$
$$\Gamma^2{}_{1\alpha} = (a^\dagger{}_{2\alpha}a^\dagger{}_{2\alpha} + a_{2\alpha}a_{2\alpha})/2 \quad (4.31)$$

We also define the Hermitian operators

$$\Gamma^2{}_{4\alpha} = (a^\dagger{}_{1\alpha}a^\dagger{}_{1\alpha} + a_{1\alpha}a_{1\alpha})/2 \quad (4.32)$$

and

$$\Gamma^2{}_{5\alpha} = i(a^\dagger{}_{2\alpha}a^\dagger{}_{2\alpha} - a_{2\alpha}a_{2\alpha})/2 \quad (4.33)$$
$$\Gamma^2{}_{6\alpha} = i(a^\dagger{}_{1\alpha}a^\dagger{}_{1\alpha} - a_{1\alpha}a_{1\alpha})/2 \quad (4.34)$$

They transform $a_{1\alpha}$ and $a^\dagger{}_{1\alpha}$ to have additional terms of the form $c_\alpha a_{2\alpha} + c_\alpha' a^\dagger{}_{2\alpha}$ where c_α and c_α' are constants; and they transform $a_{2\alpha}$ and $a^\dagger{}_{2\alpha}$ to have additional terms of the form $c_\alpha a_{1\alpha} + c_\alpha' a^\dagger{}_{1\alpha}$.

The complete set of Γ operators are part of an su(1,1) group. This group appears in Twistor Theory. A relation/unification of the group of these Bogoliubov transformations and Twistor Theory may eventually emerge.

The transformation $B_{2\alpha}B_{1\alpha}$ generates the maps of eqs. 4.21 and 4.22. As we will see next the key terms for the transition between annihilation/creation operators of two coordinate systems, A and B, have the form:

$$[c'_{i1}a^\dagger{}_{1\alpha} + c'_{i2}a^\dagger{}_{2\alpha} + C'_{i1}a_{1\alpha} + C'_{i2}a_{2\alpha}] \quad (4.35)$$

if we construct states based on $a^\dagger{}_{2\beta}$ and $a^\dagger{}_{2\alpha}$ with the vacuum $|0>_2$. Note there are four constants, which are functions of α, in eq. 4.35. These constants are determined by the four parameters in $B_{2\alpha}B_{1\alpha}$. Thus the transformation between coordinate systems embodied in $B_{2\alpha}B_{1\alpha}$ determines the constants in eq. 4.35.

We now define a combined General Relativistic – Internal Symmetry (GRIS) form of transformation that causes rotations of the components of a dimension array. Note that dimension arrays can be viewed as abstractions from the algebra of creation and annihilation operators. So the GRIS transformations can be applied to dimension arrays.

The GRIS transformations have two forms. One form applies directly to the dimension arrays treated as matrices. We chose a different form by mapping the dimension array to a vector of 2^{r+4} components. We define a Fundamental Reference Frames (FRF) for the 2^{r+4} component vector. We call it a reference frame because we can apply GRIS transformations to it to obtain a different FRF.

Now after defining the FRFs we can now inquire if there is a space corresponding to it. The answer is yes – there is a new Cosmos space, called a HyperCosmos HyperUnification Space, with dimension

$$r' = 2r + 4 \quad (8.5)$$

based on the number of its square dimension array components:

$$2^{2r+8} = 2^{r'+4}$$

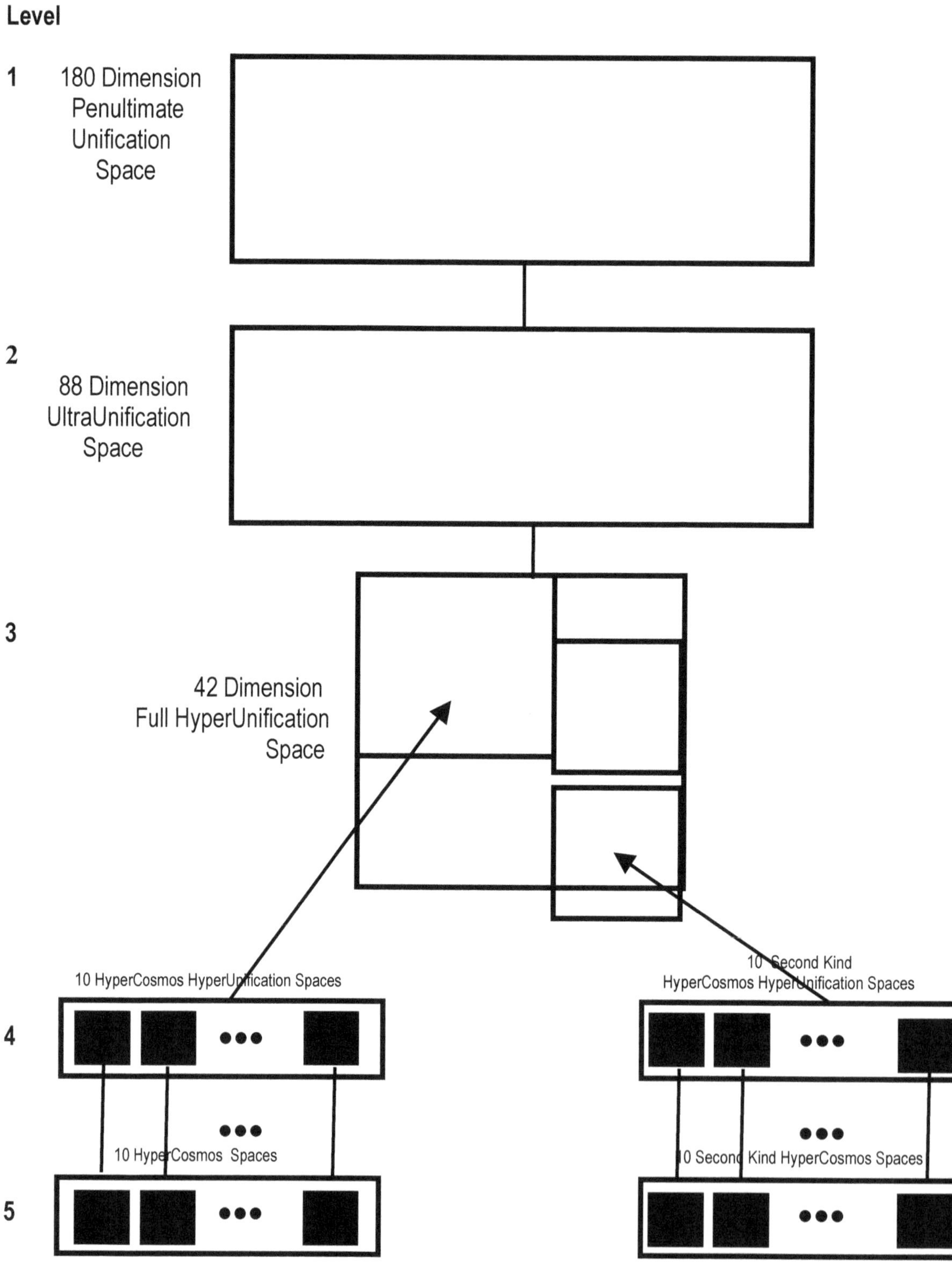

Figure 8.10. Diagram of the five levels of the Cosmos. They contain 43 spaces.

So there are ten HyperCosmos HyperUnification Spaces – one space for each of the 10 HyperCosmos spaces. They appear in Fig. 8.10 together with 10 Second Kind HyperCosmos Spaces which are generated by a similar approach.

As a result all 20 spaces have HyperUnification Spaces that implement rotations of fermions and internal symmetries.

8.4.2 The 42 Dimension Full HyperUnification Space

We now consider the combined unification of all or some of the HyperCosmos spaces in a *Full HyperUnification Space* (a 42 space-time dimension space). Each space has a dimension array with d_{dN} elements. This array is expressed as a vector v_N with d_{dN} components residing in the space's unification space with space-time dimension r' given by eq. 8.5.

Now consider the sum of all ten HyperUnification spaces vectors:

$$v_S = \sum_{N=0}^{9} v_N \tag{8.6}$$

$$= 4^2 + 8^2 + 16^2 + 32^2 + 64^2 + 128^2 + 256^2 + 512^2 + 1024^2 + 2048^2$$

with

$$v_N = d_{dN} \tag{8.7}$$

These elements form a v_S-vector in the Full HyperUnification space. The sum of the ten vectors of the ten HyperCosmos HyperUnification spaces vectors is a vector d_{cS} of the Full HyperUnification Space:

$$d_{cS} = v_S = 5{,}592{,}400 \tag{8.8}$$

There is a corresponding set of square dimension array blocks whose total is

$$d_{dS} = 4^4 + 8^4 + 16^4 + 32^4 + 64^4 + 128^4 + 256^4\text{^}+ 512^4 + 1024^4 + 2048^4$$
$$= 1.8765\times10^{13} \tag{8.9}$$

The Full HyperUnification Space plays the role of unification for all ten HyperCosmos spaces. Each of the ten HyperCosmos HyperUnification space vectors of size d_{dN} in v_S individually has d_{dN} components. Correspondingly, each subvector has a $d_{dN}\times d_{dN}$ square array along the diagonal of a combined transformation which we call a *HyperUnification Transformation.* (Fig. 8.11)

The dimension array block size defined by the d_{cS} vector is

$$d_H = 5{,}592{,}400^2 = 3.12749\times10^{13} \tag{8.10}$$

If we treat this square array d_H as corresponding to a HyperCosmos space then we can calculate the corresponding[19] space-time dimension r_H

---

[19] This is the dimension implied by d_H.

$$r_H = \log_2(d_H/16) \qquad (8.11)$$
$$= 2\log_2(5{,}592{,}400) - 4 = 40.83$$

using

$$d_{dH} = 2^{22-2N_H} = 2^{r_H+4} \qquad (8.12)$$

The dimension r_H is non-integral. Noting that space-time dimensions are even numbered in HyperCosmos spaces. we *define* the *Full HyperUnification Space* to have

$$r_H = 42 \qquad (8.13)$$

with (N_H is the Blaha number in Fig. 7.1)

$$N_H = \tfrac{1}{2}(18 - r_H) = -12$$

space-time dimensions.[20] It is the smallest space-time dimension that has a dimension array block containing the d_H dimension array block as shown in Fig. 8.12. (Note: the dimension array block d_{dN_H} has the same size as the General Relativistic transformation between the FRF and a static space-time.)

8.4.3 The General Relativistic Transformation

The General Relativistic transformations of the $r_H = 42$ Full HyperUnification Space includes the General Relativistic transformations of any or all, HyperUnification spaces along its "diagonal" where blocks correspond to individual unification space General Relativistic transformations. See Fig. 8.11. It also includes much more in the form of arbitrary, not necessarily block diagonal, transformations that mix various HyperCosmos unification spaces.

The 42 space-time dimension array has the vector column size

$$d_{cN_H} = 2^{42/2+2} = 2^{23} = 8{,}388{,}608 \qquad (8.14a)$$

with the number of array elements being

$$d_{dN_H} = 8{,}388{,}608^2 = 7.03687\times10^{13} \qquad (8.14b)$$

Since d_{cN_H} is larger than d_{cS} there is an apparent excess in the 42 space-time dimension Full HyperUnification Space which we consider below. (See Fig. 8.12.)

We can define a set of HyperUnification Space General Relativistic transformations that transform in each of the individual HyperUnification sub-blocks separately. We can also define transformations that transform combinations of all 10 HyperUnification spaces including the entire set of HyperUnification spaces.

---

[20] Negative values of N are allowed to specify Cayley number type dimension arrays.

8.4.4 Full HyperUnification Space FRF

The FRF vector for the Full HyperUnification Space has d_{cN_H} components. Comparing it to the size of d_{cS} = 5,592,400-vector we find the General Relativistic transformations of the 10 HyperCosmos unification spaces are a subset of the 42 space-time dimension HyperUnification Space transformations. The column excess is 2,796,208 = 2,796,200 + 8 dimensions. There is more in the figure.

The ratio of the larger box labeled A to the smaller box labeled B in Fig. 8.12 is exactly

$$d_{cN_H}/d_{cS} = 1.5 = 3/2 \tag{8.15}$$

Thus the B box column size d_{cB} satisfies

$$d_{cB} = \tfrac{1}{2}\, d_{cS} = 2{,}796{,}200 \tag{8.16}$$

and

$$d_{cB} = \tfrac{1}{2} \sum_{N=0}^{9} v_N \tag{8.17}$$

8.4.5 Second Kind HyperCosmos in 42 Dimension Space

*The Full HyperUnification space holds the 10 HyperCosmos HyperUnification spaces **plus the 10 Second Kind HyperCosmos HyperUnification spaces.*** The minimal space[21] that will hold the 10 HyperCosmos HyperUnification spaces has 42 space-time dimensions. (We require the Full HyperUnification space to have the same form as the HyperCosmos spaces but with a higher space-time dimension.)

The 42 space-time dimension Full HyperUnification space dimension array holds both the 10 HyperCosmos HyperUnification spaces and the 10 Second Kind HyperCosmos HyperUnification spaces dimension arrays (and the 42 space-time dimension General Relativistic transformations).

The Full HyperUnification space total dimensions expressed as a square array has 8,388,608 by 8,388,608 components. See Fig. 8.12. The 10 HyperCosmos HyperUnification spaces' dimension arrays occupy the 5,592,400 by 5,592,400 square array components block labeled by "A". The 10 Second Kind HyperCosmos HyperUnification spaces' dimension arrays occupy the 2,796,200 by 2,796,200 square array components block labeled by "B". The remaining 8 by 8 square array block labeled by "C" complete the Full HyperUnification space dimension array.[22]

For both A and B blocks, the HyperUnification space dimension blocks within them are strung along the diagonal as shown in Fig. 8.12.

Note that the length of block B, 2,796,200, is one half of the length of block A, 5,592,400. The difference is due to block A having both a 2,796,200 by 2,796,200 subblock for its Normal Matter sector and a 2,796,200 by 2,796,200 subblock for its Dark Matter sector. Block B only has a Normal Matter sector. Thus we see that the 42 space-time dimension array has three parts. This fact will be of interest later.

---

[21] A space with smaller than 42 space-time dimensions will not hold the dimension arrays of the 10 HyperCosmos HyperUnification spaces.

[22] See Blaha (2023b).

The 42 space exhibits a 1, 4, 9 sequence of blocks of dimensions. They are the squares of the 1, 2, 3 sequence of the A and B block sides and the total 42 space dimension array block side 2,796,200, 5,592,400, 8,388,608.

The B block holds the HyperUnification spaces of the 10 Second Kind HyperCosmos spaces.

8.4.6 Second Kind HyperCosmos Dark Matter Anyons?

The C block has 64 dimensions in an 8 × 8 square array with sides of 8 dimensions. This block corresponds to an r = 0 Second Kind HyperCosmos dimension array of size 2 × 4 = 8 dimensions. We see that there is an r = 0 term in block B. Thus block C duplicates that term.

We may view this r = 0 duplicate as a specification of a miniverse that could exist in our four dimension space. This miniverse could have eight[23] fermions in two sets of four Normal fermions (no Dark fermions) following the spectrum structure described earlier. They may each be SU(4) quark-lepton quadruplets[24]: one up-type; one down-type. Possible symmetry groups include: SU(4) and SU(2)⊗SU(2)⊗U(1) and $SO^+(1, 3)$⊗SU(2)⊗U(1).

If r = 0 miniverses exist in our universe in quantity then they could be candidates for Dark Matter. They only have the gravitational interaction outside of their miniverses. Since the space-time dimension is zero, their nature as fermions is clouded. They may support anticommutation relations; they may support commutation relations; they may support some intermediate form of commutation relations. Consequently these miniverses are a possible form of *anyon.*

The third possibility above opens the door to a space-time internal to a miniverse. The space-time of a PseudoQuantum wave function describes the PseudoQuantum universe in a child space. The PseudoQuantum's universe's dimension array contains a set of dimensions for the internal space-time of a universe. The universe's internal space-time supports the wave functions of the particles and energies within the universe. The $SO^+(1, 3)$ group above is for the case of a miniverse with an internal space-time. In this case the minverse may have an internal dynamical evolution.

The absence of any experimental indication of Dark matter other than gravitation supports the possibility of an anyon Dark matter interpretation.

8.4.7 Impact of Unification up to 42 Dimension Full HyperUnification Space

The preceding discussions in section 8.4 led to this 42 dimension space. We showed it contained the 10 HyperCosmos spaces, the 10 Second Kind HyperCosmos spaces and an 8×8 square array. This space supports rotations amongst the 20 spaces. It can be viewed as GRIS rotations of their irreducible representations of symmetries or as GRIS rotations among the fermions of the 20 spaces. Thus we have a new unification of the Cosmos spaces.

---

[23] The size of the miniverse dimension array.

[24] Possibly broken to SU(3)⊗U(1).

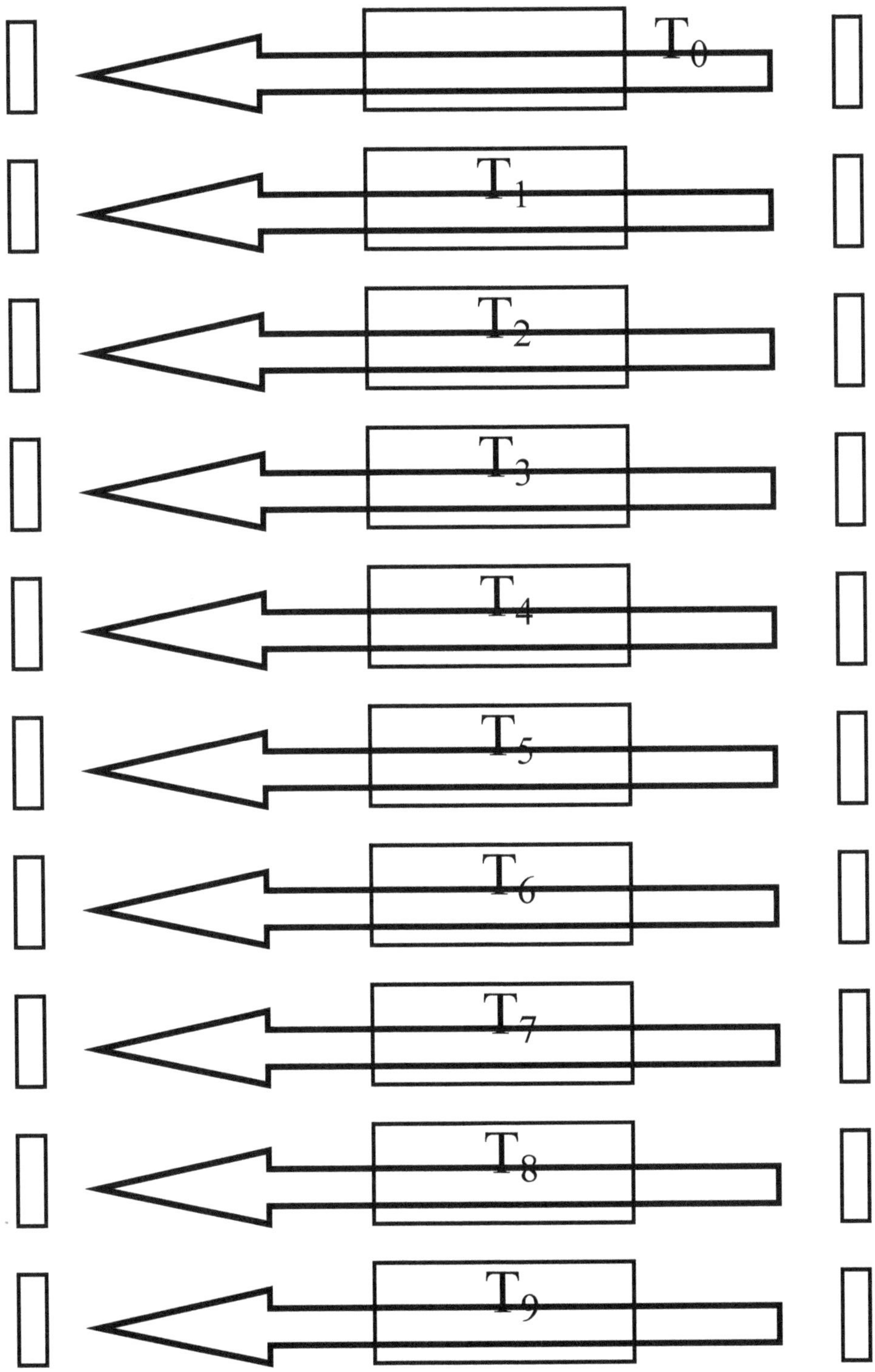

Figure 8.11. HyperCosmos transformations in the Full HyperUnification Space.

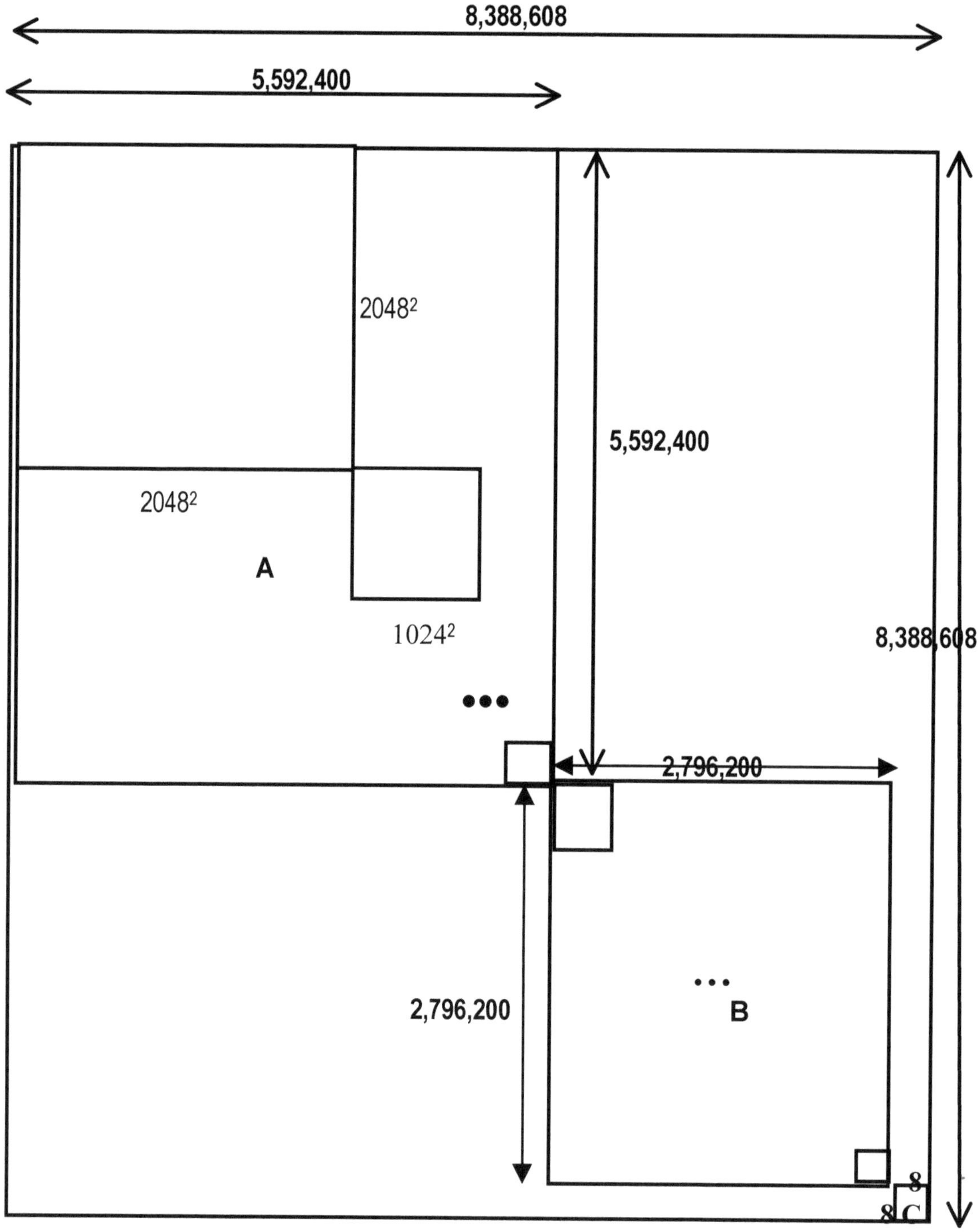

Figure 8.12. Form of the 42 space-time dimnesion of the Full HyperUnification space. All blocks are square. The figure is not drawn to scale.

8.4.6 UltraUnification Space – 88 Dimension Space

The 42 dimension Full HyperUnification Space has a unification space of 88 dimensions. This follows from eq. 8.5, which gives 88 = 2×42 + 4. We called the 88 dimension space the *UltraUnification Space*. See Fig. 8.10. The role of this space is to support unification within the 42 dimension space.

The 88 dimension UltraUnification Space supports GRIS rotations amongst the 20 spaces embedded within the 42 dimension Full HyperUnification Space. It can be viewed as GRIS rotations of their irreducible representations of the Full HyperUnification Space's set of 42 dimension symmetries or as GRIS rotations among the fermions of the 20 spaces embedded within the 42 dimension Full HyperUnification Space. Thus we have further unification of the Cosmos spaces embedded within the Full HyperUnification Space.

8.4.7 Beyond UltraUnification Space – 180 Dimension Penultimate Unification Space

The 88 dimension UltraUnification space has a unification space that we call the *Penultimate Unification Space*. This follows from eq. 8.5, which gives 180 = 2×88 + 4.

The 180 dimension Penultimate Unification Space supports GRIS rotations amongst the contents of the 88 dimension UltraUnification Space. It can be viewed as GRIS rotations of the symmetry irreducible representations of the UltraUnification Space's set of symmetries or as GRIS rotations among the fermions of the UltraUnification Space's Thus we have another unification of the Cosmos spaces.

One purpose of this unification is to support a new view of the unification processes as a sequence of unifications that reduces the contents of the Cosmos spaces to GRIS rotations of a subset of space contents (dimensions) found in a Fundamental Reference Frame (FRF). This subset can be taken to be a single dimension (quantity) as we discuss in the following chapter.

A more important purpose is to reduce the set of transformations of Cosmos spaces from GRIS transformations to strictly General Relativistic (GR) transformations. We saw the set of dimensions (coordinates) in the 42 dimension Full HyperUnification Space numbered 2^{46} = 8,388,608×8,388,608 by eq. 8.14a. The set of dimensions in the dimension array vector of the 180 dimension Penultimate Unification Space is 2^{92}. Thus the vector separates nicely into 2^{46} Full HyperUnification Space part and another non-Full HyperUnification Space part. The GR transformations of the 180 dimension space can be separated into a GR part for one $2^{46}\times2^{46}$ set of coordinates and a separate GR part for the other $2^{46}\times2^{46}$ set of coordinates. We partition the set of GRIS transformations of Penultimate Unification Space into a GR factor for the rotations of the 42 dimension Full HyperUnification Space and another set of GR transformations for the other set of coordinates. *Symbolically* we can represent the set of Penultimate transformation as $GR_{42}\otimes GR_{rest}$. The Penultimate GR_{42} set of transformations embody the GRIS transformations, both GR and Internal Symmetry transformations, of the 42 Dimension Full HyperUnification Space.

Thus Penultimate Unification Space's strictly GR transformations contain the GRIS transformations (including symmetry group transformations) of the Full HyperUnification Space.

8.5 A Sequence of Unification Spaces

We now have four levels of unification: HyperCosmos HyperUnification Space with r = 18, Full HyperUnification Space, UltraUnification Space, and Penultimate Unification Space with dimensions: 18, 42, 88, 180 respectively. We choose the dimension for the HyperCosmos HyperUnification Spaces to be 18 since this space can "hold" all nine spaces below it and thus constitutes a type of unification space.

The dimensions form a sequence that has approximately a multiplier factor of 2 due to eq. 8.5.

Space	r	
HyperCosmos HyperUnification Space	**18**	
Multiplier		2.33
Full HyperUnification Space	**42**	
Multiplier		2.1
UltraUnification Space	**88**	
Multiplier		2.05
Penultimate Unification Space	**180**	

Figure 8.13 The sequence of unification spaces.

We shall discuss this sequence in chapter 10.

9. The Seeds of the Cosmos and Fundamental Reference Frames (FRF)

This chapter shows how the Cosmos may be viewed as generated from simple features – ingredients that might be viewed as seeds. Just as organisms grow from seeds the Cosmos emerges from its seeds in a mathematical manner.

We begin by considering the 'seeds" and indicate the pattern of growth that follows. Then we discuss Fundamental Reference Frames (FRF) that establish a path to the features of universes: particles, symmetry groups and interactions.

9.1 The Seeds of the Cosmos

The primary seed of the Cosmos is the mathematical point that grows through a series of steps to a fractal curve and thence to a two dimension space. This space is the basis of the Master Equation of Physical Reality. From the Master Equation, and the features of totally antisymmetric tensors and fermion quantum fields, we developed the structure of the Cosmos spaces. The forty three spaces, that we found, are listed in Fig. 8.10. We see the beginning of the Cosmos organism.

Another seed is the structure of the unification levels. We will see that at each of the levels it is possible to transform the set of dimensions of each dimension array to a single point in a "reference frame" that we have called "The Rest Frame of the Array."

9.2 Rest Frame of a Dimension Array

In Relativistic Mechanics we are familiar with the rest frame of a particle. In that frame the spatial 3-momentum of a particle is zero. Only the energy is not zero – it is equal to the mass of the particle.

In chapter 8 we considered GR and GRIS transformations for dimension arrays. It was convenient to map a dimension array to a vector and then to apply GR and GRIS transformations to it. This process was analogous to the application of GR transformations to a fermion wave function Fourier expansion as we showed in our papers of the 1970s. The creation and annihilation operators were "rotated." This effect is well known to occur in transformations to accelerated reference frames.

Since dimension arrays can be viewed as abstracted from fermion creation and annihilation operators, their components will undergo similar transformations under GR and GRIS transformations.

The analogy to relativistic Mechanics leads us to define Reference Frames for dimension arrays. We defined a type of reference frame that we called a *Fundamental Reference Frame* (FRF) in earlier books such as Blaha (2022c). We now turn to their discussion.

9.3 Fundamental Reference Frame for Our r = 4 Universe

An examination of a dimension array for the cases of the fermion spectrum and the set of internal symmetries reveals an apparent duplication of features. For example, in the case of our r = 4 space and universe Fig. 8.3 shows the fermion spectrum can be viewed as 16 duplicates of 16 fermions consisting of 8 Normal fundamental fermions and 8 Dark fundamental fermions. Fig. 8.2 shows the internal symmetries have 16 duplicates of 16 real-valued dimension sets of U(4)⊗U(4) groups; The groups, after symmetry breakdowns, have the forms: SU(4) (or SU(3)⊗U(1)), SU(2)⊗U(1)⊗SL(2, C), SU(2)⊗U(1)⊗U(2), Generation U(4), and Layer U(4). *Thus we see 16 duplicates of 16 dimension parts in the r = 4 dimension array.*

In chapter 9 of Blaha (2022c) (and in preceding books) we showed that one could define GRIS transformations for our r = 4 space that can transform a vector composed of the dimension array components. We now consider the case of defining a vector composed of 16 non-zero dimensions and 240 zero dimensions.[25] We call this vector the r = 4 *Fundamental Reference Frame (FRF)*. We define a GRIS transformation that maps this FRF, denoted f_1, to a full non-zero set of dimensions that constitutes the dimension array when transformed back to a square array. Similarly we define a GRIS transformation that maps a corresponding f_1 FRF for fermions to the full dimension array spectrum of 256 fermions. Also we define a GRIS transformation that maps a corresponding f_1 FRF for internal symmetry dimensions to the full dimension array set of internal symmetry dimensions.

We thus have a method for reducing a dimension array to an FRF rather like a coordinate transformation that reduces a particle to its rest frame. As we will see, this method leads to a method for reducing sequences to a more general, yet simpler form, in chapter 10.

9.4 Fundamental Reference Frame for Larger Spaces and Universes

The spaces of Cosmos Theory have dimension arrays that quadruple in size as the dimension r increases by 2. See d_{dr} column in Fig. 7.1. Consequently we can view the dimension array for r + 2 as composed of four quadrants – each consisting of a dimension array of dimension r. We therefore can directly extrapolate the results of section 9.2.

We begin in dimension r with a dimension array with the size 2^{r+4} we define an FRF vector f_r that has $2^{r/2+2}$ non-zero components and $2^{r+4} - 2^{r/2+2}$ zero-value components.

Then we can use GRIS transformations to populate the dimension array. We can also use GRIS transformations for a fermion f_r to populate the fermion dimension array, and GRIS transformations for internal symmetries dimensions f_r to populate the internal symmetries dimension array.

---

[25] As we have said many times previously, the value of a dimension component in a dimension array is irrelevant except in distinguishing zero components from non-zero components.

9.5 Relating FRF's Using the Quadrupling Feature

We now relate FRF's of differing dimensions r′ > r using quadrupling. Let

$$r' = r + 2n \tag{9.1}$$

where n is an integer. Then the sizes of their dimension arrays satisfy

$$2^{r'+4} = 2^{2n}2^{r+4} \tag{9.2}$$

as does the number of $f_{r'}$ FRF vector components. Eq. 9.2 shows that there are $2^{2n} = 4^n$ iterations of the r dimension array in the r′ dimension array. The r FRF has $2^{r/2+2}$ non-zero components and $2^{r+4} - 2^{r/2+2}$ zero-valued components. The r′ FRF has $2^{r'/2+2}$ non-zero components and $2^{r'+4} - 2^{r'/2+2}$ zero-valued components.

The 2^{2n} iterations of the array size implies the r′ FRF has

$$2^n 2^{r/2+2} = 2^{(r + 2n)/2+2} = 2^{r'/2+2} \tag{9.3}$$

iterations. So there are 2^n iterations of the r FRF's non-zero components and $2^{r'+4} - 2^n 2^{r/2+2}$ zero-valued components in the r′ FRF. We have thus represented the r′ FRF in terms of the r FRF. See Fig. 9.1 for a list of FRF non-zero components.

In consequence, we see the r′ space dimension array duplicates (actually multiply quadruplicates) the r space dimension array. We can take the r′ fermions and internal symmetries to be replicates of the r fermions and symmetries.

For example, we can express the contents of the r′ dimension FRF in terms of the number of repetitions of the $2^4 = 16$ non-zero components of the r = 4 FRF as

$$2^n\, 2^4 = 2^{r'/2 + 2}$$

where r′ = 4 + 2n or n = (r′ – 4)2. For r′ = 6 we see $2^n\, 2^4 = 2^5$. See Fig. 9.1 below.

When we consider the sequences of coupling constants and fermion masses, we can assume higher dimension versions may well have similar sequences to those of our space and UST universe. The reader may wish to consider Blaha (2022e) which provides additional detail.

9.6 Reduction of an FRF to a Single Dimension PFRF

The above discussions give us a reduction to FRFs with numerous non-zero components. We can take a dimension array (and its associated fermions and internal symmetries) and reduce it significantly with GRIS transformations.

We now consider an extension of this reduction process – reducing a dimension array to a vector with only one non-zero component. We call these representations Point Fundamental Reference Frames (PFRF). Our purpose is clear. We wish to reduce the Coamos Theory dimension arrays (and thus their dimensions) using GRIS transformations to a single dimension. With this capability we can view Cosmos Theory as emanating from a "seed" in a conceptual growth process. However we must continue to realize that our description, based on a series of processes, is conceptual since the Cosmos Theory structure is ever present without beginning or end.

Method: Express each dimension array as a vector. Find a GRIS transformation that maps all components of the vector to zero except for one component, which becomes the seed. From the seed, apply GRIS transformations that generate the desired dimension array. The seed is the non-zero dimension. But it can be taken to be an "ancestor" fermion or an "ancestor" dimension for an internal symmetry.

9.7 PFRF for HyperCosmos HyperUnification Spaces, the Full HyperUnification Space, the UltraUnification Space, and the Penultimate Unification Space

These spaces can also have their dimension arrays reduced to an FRF vector with only one non-zero dimension following a process similar to that of section 9.6. We can construct each of the 43 spaces separately from 43 non-zero dimensions – one non-zero dimension for each space.

Cosmos Theory also supports sequential construction of spaces. In the extreme, and most interesting case, we can define an FRF for the 180 dimension Penultimate Unification Space with one non-zero dimension in its FRF's vector. We can GR transform this vector to the full dimension array of the 180 dimension. Then we can extract the dimension array of the 88 dimension space; then further extract the dimension array of the 42 dimension space.

From the 42 dimension space we can extract the 10 HyperCosmos HyperUnification Spaces, the 10 HyperCosmos spaces of the Second Kind HyperUnification Spaces, the 10 HyperCosmos Spaces, and the 10 HyperCosmos spaces of the Second Kind Spaces. See Fig. 8.10 for the path to the full construction of the dimension arrays of all spaces from one non-zero dimension in the Penultimate Unification Space. The purpose of this process is to establish the possibility of a descent of all spaces from one (non-zero) dimension.

9.8 Threads of the Cosmos – Physical Reality

We now can see a conceptual evolutionary process (that does not happen in time) with the threads of Reality:

1. A mathematical point from which the Master Equation of the Cosmos arises.

2. A PFRF for each Cosmos space that acts as the seed for its dimension array. All that is needed is one non-zero dimension in Penultimate Unification Space.

3. A point (fundamental value) for each Elementary Particle entity from which its sequences emerge: Cosmos spaces sequence, coupling constants, up-type fermion masses, down-type fermion masses, sequence of unification spaces, and the sequence of sequences. (See chapter 10.)

We see the Physical Reality of Sequence Theory and Cosmos Theory can be based on "emanations" from points of a very few threads.

Contents of Fundamental Reference Frames

r	Number of Physical Elements nonzero dimensions	Total Number of Elements
18	2^{11}	2^{22}
16	2^{10}	2^{20}
14	2^{9}	2^{18}
12	2^{8}	2^{16}
10	2^{7}	2^{14}
8	2^{6}	2^{12}
6	2^{5}	2^{10}
4	2^{4}	2^{8}
2	2^{3}	2^{6}
0	2^{2}	2^{4}
r	$2^{2/2+2}$	2^{2+4}

Figure 9.1. The non-zero components of the ten types of Fundamental Reference Frames for the ten spaces. The elements are fermions or dimensions or internal symmetry fundamental representation dimensions.

10. Sequences and Fermion Strata

We developed Sequence Theory[26] in chapter 7. Fig. 7.2 has a table of sequences for Sequence Theory and Cosmos Theory. We now must add a new row **A2** to this table in view of the sequence of unification spaces exhibited in Fig. 8.13. The result is Fig. 10.1.

Sequence	n_φ	g_g multiplier factor	λ_{n_φ}
A1 Cosmos spaces dimension array column lengths $2^{r/2+2}$	4	2	20.79
A2 Cosmos unification space dimension r	4	2	20.79
B Coupling Constants Multiplier	2	4	5.39
C Fermion Sequence 2 Masses	1	$2/3 \times 2^5$	1.3
D Fermion Sequence 1 Masses	$½^{½}$	$3/2 \times \pi 2^5$	0.65
E n_φ Values Sequence; A Sequence of Sequences	½	95939	0.325

Figure 10.1. Cosmos Theory Sequences. Note: The fermion Mass sequences must apply to mass sequences in other generations and other layers up to overall factors. The coupling constant sequence must apply to the coupling constant sequences in other layers up to overall factors. The values of n_φ, g_g, and λ_{n_φ} should apply for all generations and layers. The sole difference is the values of the starting sequence values $c_{n\varphi n}$ from eq. 7.2a. Note $\lambda_{n_\varphi} > 0.25$ is required by eq. 7.1 in order to have the bound state energy spectrum. Thus the possible additional line with $n_\varphi = ¼$ is ruled out. Note the unification space dimension multiplier is approximated by 2. See Fig. 8.13 for the exact multiples. Fig. 11.5. will extend the sequences with a new sequence determined **by setting λ = .303 with multiplier 844242.**

We now consider the meaning of the sequences in Fig. 10.1. First note the first line A1 applies to the structure of Cosmos Theory in general. The second line A2 applies to the structure of the unification spaces of Cosmos Theory. The remaining lines are for individual universes. We view these lines as applicable in all universes of all ten Cosmos spaces. The features of the individual sequences are:

A1 Cosmos dimension array column sizes ($2^{r/2+2}$) Multiplier $n_\varphi = 4$
Cosmos spaces' dimension array columns grow by factors of 2 as the dimension increases by 2. This sequence directly matches the sequence of line segments in the Hilbert fractal curve construction as well as the sequence of line segments in the fractal Cosmos Curve construction.

A2 The approximate dimension r multiplier $g_g = 2$ of Unification spaces. Note $n_\varphi = 4$. Cosmos unification spaces dimensions grow by factors of 2.

---

[26] See Blaha (2025b).

B Coupling Constants Multiplier $n_\varphi = 2$
The coupling constants α values grow by factors of 4. See chapter 11.

C Fermion Sequence 2 Masses $n_\varphi = 1$
The sequence 2 fermions grow by factors of 32. See chapter 11.

D Fermion Sequence 1 Masses $n_\varphi = \frac{1}{2}^{\frac{1}{2}} = 2^{\frac{1}{2}}/2 = 0.707107$
The sequence 1 fermions grow by factors of 32π. See chapter 11. This n_φ value has a multiplier of $2^{\frac{1}{2}}$ from the $n_\varphi = 1$ value and also to the $n_\varphi = \frac{1}{2}$ value. Thus it is mid-sequence. It has the value ½ times $2^{\frac{1}{2}}$ - multiplicatively greater than the sequence value of the next $n_\varphi = \frac{1}{2}$ value.

E Sequence of n_φ Values $n_\varphi = 1/2$
This is the sequence of n_φ values in Fig. 10.1. The items in line E are:

$$\begin{aligned} \lambda_{n\varphi=\frac{1}{2}} &= n_\varphi^2/m = \tfrac{1}{4}/m = 0.325 \\ g_{g\frac{1}{2}} &= \exp(\pi/(\lambda_{n\varphi=\frac{1}{2}} - \tfrac{1}{4})^{\frac{1}{2}}) \\ &= 95939 \end{aligned} \tag{10.2}$$

The sequence of $n_\varphi = \frac{1}{2}$ values has a growth multiplier $g_{g\frac{1}{2}} = 95939$. This extremely large value is, at first glance, somewhat strange. It is much larger than the analogous multipliers for fermions in items C and D. However, the relation of the (changed[27]) ν' and ν_e Neutrino masses revealed in Fig. 10.2 reflects our belief that there are some new parts to the complete spectrum of fundamental fermions. Previously we used the mass values[28]

$$\nu' = \nu_e\, e^2(\alpha_0^2)^{-1} \approx \pi e^2 g_{g\frac{1}{2}} \nu_e \tag{10.3}$$

where

$$g_{g\frac{1}{2}} = 95939 \approx 1/(\pi\alpha_0^2) = 97852 \tag{10.4}$$

(from Fig. 10.1) using

$$\alpha_0 = e^2/4096 = 0.0018036$$

We now use a ν' mass based on the mass multiplier of 95939:

$$\nu' = 95939\ \nu_e \tag{10.3a}$$

The treatment of 95939 as a multiplier for mass is consistent with the C and D items which also give multipliers for mass sequences of fundamental fermions. *We use 95939 tentatively as the mass multiplier realizing it is subject to change.*

---

[27] Our previous choice of υ′ mass was an estimate based on fermion mass sequences.
[28] The choice of the ν' mass was made in our earlier book. It was based on making sequence 2 masses approximately follow a power sequence form, which was not possible if we used the ν_e mass.

Having introduced a heavier electron neutrino mass we assume there are corresponding heavier muon and tau particles. The known neutrinos appear to be "separate" from the heavier neutrinos. We therefore have suggested that the known neutrinos have a set of heavier neutrino counterparts.

In our UST formulation of the fermion spectrum the number of low mass fermions is 256. It is comprised of four layers of four generations of 16 Normal and Dark fundamental fermions. We call this set of fermions *stratum* 1. The Normal neutrinos of stratum 1 are known with the exception of $\upsilon_?$. The charged fermions (both Normal and Dark) of stratum 1 are unknown at present.

The heavier neutrinos are in stratum 2. They are not known at present. The neutrinos of this set are the "primed" neutrinos. We assume they have charged fermion partners in stratum 2. Their 7 Normal charged partners of stratum 2, generation 1, layer 1 are the charged fermions (quarks and leptons) with which we are familiar. The remaining charged fermions are not known at present.

The total number of fermions in stratum 2 in the UST is 256. This set of fermions consists of four layers of four generations of 16 Normal and Dark fundamental fermions. This set of fermions constitutes *stratum* 2.

We can summarize the first two strata with:

Stratum 1

υ_e known	7 Charged fermions	Dark υ_e	7 Dark Charged fermions
υ_μ known	7 Charged fermions	Dark υ_μ	7 Dark Charged fermions
υ_τ known	7 Charged fermions	Dark υ_τ	7 Dark Charged fermions
$\upsilon_?$	7 Charged fermions	Dark $\upsilon_?$	7 Dark Charged fermions

Stratum 2

υ_e'	7 Charged fermions known	Dark υ_e	7 Dark Charged fermions
υ_μ'	7 Charged fermions	Dark υ_μ	7 Dark Charged fermions
υ_τ'	7 Charged fermions	Dark υ_τ	7 Dark Charged fermions
$\upsilon_?'$	7 Charged fermions	Dark $\upsilon_?$	7 Dark Charged fermions

Figure 10.2. First two strata with known fermions marked.

10.1 Four Strata of Fundamental Fermions

We have seen the beginning of the development of the concept of strata in the E row multiplier of 95939 in Fig. 10.1. We take that to be a mass multiplier of a new structure in the fermion mass spectrums in a manner analogous to the mass multipliers in rows C and D. We assume the existence of a new structuring of fermions consisting of strata. The strata are duplicates of each other on a fermion by fermion basis – the same quantum numbers, generations, and layers. They differ in fermion masses. Each fermion in stratum n has a mass that is a factor of 95939 more than the corresponding mass in stratum n – 1.

The question arises: How many strata are there? We believe the most reasonable number is four. Note the generations are numbered in 4's; the layers are numbered in 4's; and the number of fermions in a mass sequence is 4. Thus 4 strata is a reasonable choice for consistency's sake. A more interesting reason for choosing 4 strata is the growth in mass. If there were a 5th stratum, and so on, their fermion masses would put them beyond interest from an experimental viewpoint: they would be very difficult to produce, would only interact with difficulty, and would undoubtedly have an extremely short lifetime.

Thus setting the number of strata to four is thus the most reasonable choice. We number the first two strata as in Fig. 10.2 above. The four strata are depicted in Fig. 10.3.

While we have focused on the r = 4 UST set of fermions, we require universes of higher space-time dimension to have sets of fermions in four strata as well. We call the strata fermions *multiplicates* and have considered Lagrangian models for them in Blaha (2025c). All fundamental fermions in all universes are "multiplicated."

The set of strata supports an extension of the fermion Cosmos Theory dimension array. They lead to a Sequence Theory fermion dimension array of 2^{r+6} dimensions due to a factor of 2^2 for the four stratums.

stratum1
$g_{g½} = 95939$
ν_e …
ν_μ …
ν_τ …
$\nu_?$ …
Plus 3 more layers
stratum2
$g_{g½} = 95939$
ν_e' … (15 Other fermions)
ν_μ' … (15 Other fermions)
ν_τ' … (15 Other fermions)
$\nu_?'$ … (15 Other fermions)
Plus 3 more layers
stratum3
$g_{g½} = 95939$
ν_e'' …
ν_μ'' …
ν_τ'' …
$\nu_?''$ …
Plus 3 more layers
stratum4
$g_{g½} = 95939$
ν_e''' …
ν_μ''' …
ν_τ''' …
$\nu_?'''$ …
Plus 3 more layers

Figure 10.3. A view of the four UST strata of the Sequence Theory fundamental fermions showing mass multipliers. Each stratum has 256 fermions in our universe (in UST theory). Taken together the 4 strata form a 2^{r+6} set of fermions in an rth dimension universe. The 4 strata each contain 4 layers, each of which, in turn, contains 4 generations of fundamental fermions. The total for the UST is 1024 fermions.

Stratum 1 Generation 1 Layer 1 e - neutrino

$\nu_e = 3.878 \times 10^{-11}$ GeV/c^2

Our ν_e mass estimate = 0.03878 ev/c^2 is consistent with known data estimates.
Plus 7 unknown charged fermions that are counterparts to the stratum 2 charged fermions below

Stratum 2 Generation 1 Layer 1 heavy e - neutrino and 7 charged fermions

$\nu' = \mathbf{95939}\ \nu = \mathbf{3.72 \times 10^{-6}}$ **GeV/c^2**

e $= 0.511 \times 10^{-3}$ GeV/c^2 = electron mass

CURRENT FERMION MASSES

u 1.76 mev/c^2 = 1.76×10^{-3} GeV/c^2

We scale the u mass by a factor of 4/5 – again to put it on a similar footing as the other masses.
Note: the u mass is 2.2 +0.5 -0.4 × 10^{-3} GeV/c^2 The 1.76 value is consistent with the lower bound 1.8.

c 1.27 GeV/c^2

t 172.76 GeV/c^2

d 3.76×10^{-3} GeV/c^2

We scale the d mass by a factor of 4/5 – again to put it on a similar footing as the other masses.
The 3.76 value is consistent with the tentative lower bound 4.4.

s 95×10^{-3} GeV/c^2

b 4.18 GeV/c^2

Figure 10.4a. The UST Fundamental first generation current fermion masses. Each fundamental fermion symbol represents its mass.

CURRENT GAMBOL FERMION MASSES

Stratum 2 Generation 1 Layer 1 7 charged *gambol* fermions – Current Fermion Mass/844242

$e_g = 6.1 \times 10^{-10}$ GeV/c^2 = gambol electron mass

$u_g = 2.08 \times 10^{-9}$ GeV/c^2

c_g 1.5×10^{-6} GeV/c^2

t_g 2.05×10^{-4} GeV/c^2

d_g 4.45×10^{-9} GeV/c^2

s_g 1.13×10^{-7} GeV/c^2

b_g 4.95×10^{-6} GeV/c^2

Figure 10.4b. The UST Fundamental first generation, 2nd stratum current gambol fermion masses. Each fundamental current gambol fermion symbol represents its mass. We obtain the gambol multiplier factor 844242 later in section 11.8. See Fig. 11.5.

10.2 Current Masses vs. Constituent Masses

The fermion masses in Figs. 10.4a and 10.4b are *current* masses. *All references to fermion masses in this book are for current masses.*

Constituent quark fermion masses, which differ, express a quark mass as a fraction of a hadron mass. For example a proton of mass M, which consists of three quarks, attributes the mass of a u quark to M/3. Constituent quark masses are determined from the masses of the hadrons which contain them.

Based on the distinction between current and constituent masses we will analogously distinguish between *current* gambol fermion masses and *constituent* gambol fermion masses. In prior gambol-related books we used constituent gambols in calculations. For large mass quarks there is only a reassuring, small difference between current and constituent quark masses. We continue to use current gambol masses in this book. *Due to experimental and theoretical questions we view all mass results as approximate.*

10.3 Limos Strata – Gambol Strata

We have successfully developed Gambol Theory in some detail in earlier books[29] for fermion and hadron decays and interactions. Gambols are the constituents of fundamental fermions. They have a fraction of the mass and of the additive quantum numbers of the particle within which they reside. Gambols are confined at current energies.

We now introduce a representation of gambols in strata mirroring the fundamental fermions. We start by defining a shadow set of four fermion strata with the features:

1. For each stratum of fermions there is a stratum of gambol fermions.
2. Each fermion in a stratum has a corresponding gambol[30] fermion with 1/844242 of the fermion's mass and 1/844242 of the fermion's additive quantum numbers.
3. A fermion is equivalent to a corresponding set of 844242 gambol fermions. This rule is modified when the gambol fermions of a fermion are expressed in terms of the gambol fermions of a lower stratum. See section 24.2 of Blaha (2025c) for a Lagrangian Model for a fundamental fermion composed of confined gambols.

The strata of fermions is now generalized to have a fermion and its corresponding fermion in each stratum. In each stratum there is a gambol fermion for each fermion with 1/844242 of the fermion's mass and 1/844242 of the fermion's additive internal symmetry quantum numbers for each of the fermion's additive quantum numbers. See Fig. 10.5.

[29] See Blaha (2023e), *Cosmos Theory: The Sub-Particle Gambol Model*; and Blaha (2024a), *Cosmos-Universe-Particle-Gambol Theory*.

[30] **The 844242 value is derived in section 11.8.**

stratum1
Fermions
Gambol Fermions
stratum2
Fermions
Gambol Fermions ...
stratum3
Fermions
Gambol Fermions ...
stratum4
Fermions
Gambol Fermions ...

Figure 10.5. A view of the augmented four strata with each containing fermions and their gambol fermions, for each Cosmos universe.

A charged fermion can be viewed as composed of constituents consisting of gambols of the same stratum or of a combination of gambols of the same stratum and lower stratum. We will also assume stratum 1 charged fermions may be composed of stratum 1 gambols.

We note:

> A 2nd stratum mass M charged fermion has a 2nd stratum gambol of mass M/844242. The corresponding 1st stratum mass is M/95939. It has a 1st stratum gambol of mass M/(95939×844242).

10.4 Examples of a Fermion with Gambol Constituents

We now consider examples of a stratum 2 fermion's set of gambols.

10.4.1 Stratum 2 Fermion Composed Only of Stratum 2 Gambols

A stratum 2 fermion may be composed of 844242 corresponding stratum 2 gambols. Each gambol has 1/844242 of the fermion's mass. The additive internal symmetry quantum numbers of each gambol is 1/844242 of the corresponding fermion's additive internal symmetry quantum number. The gambol masses add to the fermion's mass. The gambol's additive internal symmetry quantum numbers add to the corresponding fermion's additive internal symmetry quantum number.

10.4.2 Stratum 2 Fermion Composed of A Combination of Stratum 2 and Stratum 1 Gambols

Suppose the stratum 2 fermion of mass M is composed of $N_1 < 844242$ stratum 2 gambols. Suppose further that N_2 of the fermion's gambols become stratum 1 gambols. Then

$$N_1 + N_2 = 844242 \tag{10.5}$$

The masses of the N_1 gambols are 1/844242 of the fermion's mass, and the internal symmetries are 1/844242 of the fermion's internal symmetries.

The masses of the N_2 stratum 1 gambols are 1/(844242×95939) of the fermion's mass M (1/844242 of the corresponding stratum 1 fermion's mass), and the internal symmetries are 1/844242 of the stratum 2 fermion's internal symmetries (also 1/844242 of the corresponding stratum 1 fermion's internal symmetries).

Note: For *corresponding* stratum 2 and stratum 1 fermions:

Mass of stratum 2 fermion = M
Mass of stratum 2 gambol = M/844242
Internal symmetries of stratum 2 gambols are 1/844242 of stratum 2 fermion
Mass of corresponding stratum 1 fermion = M/95939
Mass of stratum 1 gambol = M/(844242×95939)
Internal symmetries of stratum 1 gambols are 1/844242 of stratum 1 fermion
Internal symmetries of stratum 1 fermion = Internal symmetries of stratum 2 fermion

As a result

$$\text{Total mass of gambols} = M(N_1\, 1/844242 + N_2/(844242\times 95939)) \qquad (10.6)$$

This total is not M in general. It becomes the gambol scale parameter s seen in previous gambol studies. See Blaha (2023e), *Cosmos Theory: The Sub-Particle Gambol Model*; and Blaha (2024a), *Cosmos-Universe-Particle-Gambol Theory.*

The equivalent "average" gambol has mass M/s in Gambol Theory. For a stratum 2 fermion

$$s = 1/[(N_1\, 1/844242 + N_2/(844242\times 95939))] \qquad (10.7)$$

or

$$s = 1/[(N_1\, 1/844242 + (844242 - N_1)/(844242\times 95939)]$$

For many values of N_1 the second term above is negligible. Then

$$s \approx 844242/N_1 \qquad (10.8)$$

In general the gambol parameter s ranges from 844242 to 1. The values are powers of 2^n:

$$s = 1, 2, 4, 8, 16, \ldots \qquad (10.9)$$

So in the present approximation we see

$$s \approx 844242/N_1 = 2^n$$

Thus

$$n = \ln(844242/N_1) \qquad (10.10)$$

and

$$n = 0 \text{ through } \ln(844242) = 16.55 \approx 16$$
$$\text{As } N_1 \rightarrow 0 \qquad s = 1/(1/844242) = 844242 \quad \text{large}$$
$$\text{As } N_1 \rightarrow 844242 \qquad s = 1 \quad \text{large}$$

In the general case

$$1/s = N_1\, 1/844242 + (844242 - N_1)/(844242\times 95939) \qquad (10.11)$$

$$= [N_1 (1 - 1/95939) + 844242/95939]/844242$$
$$\approx (N_1 + 844242/95939)/844242 \approx 2^{-n}$$

and

$$s \approx 844242/(N_1 + 844242/95939) \quad (10.12)$$
$$N_1 \approx 1 + 844242/s - 844242/95939 = 844242/s - 7.8$$

Thus for 2nd stratum fermions the 2nd and 1st gambol strata are equivalent to gambols in Gambol Theory as seen in our earlier books.

For 3rd stratum fermions the 3rd and 2nd gambol strata are equivalent to gambols in Gambol Theory.

For 4th stratum fermions the 4th and 3rd gambol strata are equivalent to gambols in Gambol Theory. For 1st stratum fermions the 1st gambol strata are equivalent to gambols in Gambol Theory.

The above separations into gambols can penetrate into lower strata gambols. For example, 4th stratum fermions can be represented by a combination of 4th, 3rd, 2nd, and 1st gambol strata.

The gambol strata development supports a Gambol Theory. The spreads in possible distributions of gambols due to N_1 in the above example (and similarly for the cases of other strata), can be well handled by the Planckian probability distributions of gambols seen in our earlier books.

Thus Sequence Theory and Gambol Theory (based on Limos spaces) are united in a comprehensive, coordinated framework.

For more details see Blaha (2023e), *Cosmos Theory: The Sub-Particle Gambol Model*; and Blaha.(2024a), *Cosmos-Universe-Particle-Gambol Theory.*

10.5 Lagrangian Field Theory with Fermion Transformations to Gambols

Models of this type are discussed in chapter 25.

10.6 Strata of Other Spaces and Universes

The preceeding discussion of fermion strata and gambol fermion strata in particular applies to any Cosmos space and its universes. Each fermion of each universe has four strata. Each fermion is replicated into four identical corresponding fermions with the only exception being their masses. The replicated fermions' masses in the strata differ by factors of 95939.

Each charged fermion in each stratum in each space/universe also has a corresponding gambol fermion that has the mass of the fermion divided by 844242. The additive internal symmetry quantum numbers of the gambol fermion are equal to the corresponding fermion additive internal symmetry quantum numbers with each divided by 844242.

10.7 New Levels of Fermion Constituents

The set of fundamental fermions for a space/universe of dimension r began as 2^{r+4}. The replication of the fermions into strata increased the total number of fermions to

2^{r+6}. The introduction of gambol fermions for each fundamental fermion increased the total number of fermions per space/universe to 2^{r+7}.

Sequence Theory in combination with Cosmos Theory and Gambol Theory (Limos) have introduced two new levels of fundamental fermions: four strata of fermions, and a gambol fermion for each fermion in each replicated fermion stratum.

Thus the number of fundamental fermions is $2^2 = 4$ times the number of fermions in the Cosmos Theory spectrum based on the dimension array size d_{dr}. The number of gambol fermions is determined later.

The motivation for new experiment is much strengthened.

11. Coupling Constants, Mass Sequences and the 844242 Mass Multiplier

In earlier books we determined the sequence of internal symmetry coupling constants, and the two sequences of fundamental fermion masses. These quantities were part of The Standard Model and the Unified SuperStandard Theory of the author. These sequences were recognized to be the result of the Sequence Theory derived by the author from the Master Equation.

In section 11.8 we show the origin of the 844242 multiplier for charged fermions and their corresponding gambol fermions.

The sequences fit nicely in a table, Fig. 10.1 of sequences parametrized by a consistent, orderly set of n_φ values: ½, $½^{½}$, 1, 2, 4. Row B is for the Coupling Constant sequence. Rows C and D were for the fermion mass sequences.

The results that we found were:

	ESTIMATE		EXPERIMENT		
Interaction	**Expression**	$\alpha_i = g^2/4\pi$ **Value**	**g**	**Known**[31] **Value** $g^2/4\pi$	**Deviation**
U(0) α_0	$e^2/4096$	0.0018036	-	-	-
U(1) $\alpha_1 = \alpha$	$e^2/1024$	0.00721438	0.303	0.0072973525643	1.15%
SU(2) $\alpha_2 = g^2/4\pi$	$e^2/256$	0.0289	0.63	0.0316	9.3%
SU(3) $\alpha_3 = \alpha_S$	$e^2/64$	0.115	1.21	0.117	1.7%
SU(4)[32] α_4	$e^2/16$	0.462	2.4?	0.458	0.087%

Figure 11.1. Coupling constants for the UST (and the Standard Model). Note factor of four relation between adjoining α values. The coupling constant α_i multiplier was 4 as shown above.

11.1 Fractionating Groups U(1/n)

This section describes fractioning unitary groups for interactions *below* U(0).[33] Then we will determine possible coupling constant values. The U(1) group elements may be represented as:

$$e^{i\theta}$$

where θ ranges from 0 to 2π. Reducing the U(1) representation to U(½) requires a reduction in the range of coordinates. This process is accomplished by modifying the representation to

$$e^{i\theta/2}$$

[31] All coupling constant values are based on data from Particle Data Group Tables of 2022.
[32] This value is based on the "doubling trend" seen in the three known coupling constants g_i. above
[33] From Blaha (2022c).

The θ range remains 0 through 2π. Thus the resulting operators are factored by 2.

A similar procedure leads to the representation of U(1/n) where n is an even integer:

$$e^{i\theta/n} \tag{11.1}$$

Thus the groups U(1/n) support fractionation of the coordinates of their group irreducible representations.[34]

11.2 Extension of Coupling Constants to U(1/n)

The coupling constants of Fig. 11.1 can be extended to U(1/n) for n = 2, 4, 8, … .The purpose of the extension is to support the possibility of interactions between gambols within fundamental fermions. In the cases that we considered previously we treated gambols as free fermions within fundamental fermions. However, if one wishes to support probability distributions of gambols it appears necessary to have interactions between them. In our previous studies we used gambols bouncing off a confining fermion's "wall" to support a Planckian probability distribution. We wish to use gambol interactions in place of (or in addition to) a confining fermion.

We wish to have very weak interactions between gambols realizing that the Deep Inelastic lepton-nucleon parton and gambol results are consistent with free gambols. Therefore we place the gambol coupling constant values between those of U(0) and U(1). We wish to tailor the gambol coupling constants to the fractionation of the gambols. Thus the ratios correspond to powers of 2:

1/s = ½, ¼, 1/8, … 1/844242, ….

As a result we choose

Group	$\boldsymbol{\alpha_i = g_i^2/4\pi}$	**g**
U(1/s)	$\alpha_0/2^s$	
$U(1/\infty)$	0	0
…		
$U(1/95939) = \alpha_0/2^{95939}$	$4.1\times10^{-28877}\alpha_0$	1.5×10^{-28876}
…		
$U(1/8) = \alpha_0/64$	$e^2/262144$	0.000028
$U(¼) = \alpha_0/16$	$e^2/65536$	0.0001125
$U(½) = \alpha_0/4 = \alpha_{½}$	$e^2/16384$	0.00045
U(0) α_0	$e^2/4096$	0.0018036
U(1) $\alpha_1 = \alpha$	$e^2/1024$	0.00721438

Figure 11.2. U(1/s) Coupling Constants.

[34] One could extend this discussion to U(1/a) where a is a real number.

11.3 Fermion Mass Sequences

	SEQUENCE 1							SEQUENCE 2						
	e		u		c		t	ν'		d		s		b
Mass:	0.511×10^{-3}		1.80×10^{-3}		1.28		171	8.4×10^{-5}		4.24×10^{-3}		102×10^{-3}		4.34
Multiplier:		$2^5\pi$		$2^5\pi$		$2^5\pi$			32		32		32	

Figure 11.3. The fundamental fermion mass sequences with approximate multiplier factors between the masses of each sequence.[35] The sequence 1 masses show a significant variation in their $\pi 2^5 = 2^5\pi$ factor but their overall trend is approximately satisfactory. Sequence 2 masses approximate the factor of $32 = 2^5$ progression. This table was found and published in earlier work. Rows C and D of Fig. 10.1 suggest $2/3 \times 2^5$ for sequence 2 and $3/2 \times\pi 2^5$ for sequence 1. Since the fermion mass sequences are only approximate (but show the trends well), and have some significant deviations due to interaction effects, the results are compatible.

11.4 Sequence of Unification Spaces

In Fig. 8.13 we showed the sequence of unification spaces had a multiplier of 2. This sequence appeared as Row A2 in Fig. 10.1 and 11.5. This sequence (row A1) is consistent with the sequence of HyperCosmos spaces seen earlier.

The sequence of the number of Physical elements (dimensions) in Fundamental Reference Frames (Fig. 9.1) has a multiplier of 2 and thus corresponds to Row A1 in Fig. 10.1 and 11.5. See Fig. 11.5 below.

11.5 Gambol Fermion Masses

Row E in Fig. 11.5 describes the Sequence of Sequences: the n_φ Values Sequence. This sequence well describes the set of sequences of Cosmos Theory.

As a result we find a strong indication that Sequence Theory has a sound Physical basis. See Fig. 11.5 below.

11.6 Gambol Fermion Masses

We have found gambol masses to be a factor of 1/844242 times the corresponding fermion mass for each fermion in each stratum in chapter 10. The gambol masses strata corresponding to the fermions in sequences 1 and 2 listed in Fig. 11.3 must have the same multipliers as the fermions. Fig. 11.4 shows the stratum 2 gambol masses.

	SEQUENCE 1							SEQUENCE 2						
	e^g		u^g		c^g		t^g	$\nu^{g'}$		d^g		s^g		b^g
Mass×844242:	0.511×10^{-3}		1.80×10^{-3}		1.28		171	8.4×10^{-5}		4.24×10^{-3}		102×10^{-3}		4.34
Multiplier:		$2^5\pi$		$2^5\pi$		$2^5\pi$			32		32		32	

Figure 11.4. Stratum 2 fundamental gambol fermion mass sequences with approximate multiplier factors between the masses of each sequence. Fermions have the superscript g to indicate they are gambol fermions. Note multiplier 844242. Note also we have a $\nu^{g'}$ listed for completeness although neutrinos have no gambols.

---

[35] Blaha (2024g) describes the process of generating the mass sequences in detail. It shows that the similarity of the previous results and the Sequence Theory results is purely coincidental.

11.7. Cosmos Coupling Constant Values Related to Fundamental Fermion Masses

This section relates the coupling constant sequence to the fermion mass sequences.

11.7.1 The Fermion Masses Sequences

In Blaha (2024g), and above, we found that the eight known fundamental fermions of the first layer of the four dimension Cosmos Theory space and UST could be separated into two sequences of four fermions. Sequence 2 which contained the down-type fermions was approximately estimated to have a series of ratios of $2^5 = 32$ factors between the masses. Sequence 1 fermions (up-type) had less consistent factors between masses. We tentatively estimated the ratio $2^5\pi = 32\pi$ based on an approximation to the t/c fermion mass ratio and to achieve a rough parallel with the down-type fermion factor. (Fig. 11.3)

The actual mass values of sequence 1 deviated significantly from the $2^5\pi$ multiplier. We attribute the deviations to additional perturbative interaction effects on masses that are beyond this model.

11.7.2 The Fundamental Fermion Masses Model

The mass sequences, as pictured above, have a geometric character similar to the energy levels and the Cosmos spaces spectrum.

We now define a 2×2 Master Equation scaling equation for *both* sequences:

$$\begin{bmatrix} d^2/dr^2 + \lambda_1/r^2 - M_1 & 0 \\ 0 & d^2/dr^2 + \lambda_2/r^2 - M_2 \end{bmatrix} \begin{bmatrix} \varphi_1 \\ \varphi_2 \end{bmatrix} = 0 \qquad (11.2)$$

We related the multiplier of each sequence to a coupling constant as in chapter 4:

$$2^5\pi = \exp(\pi/(\lambda_1 - 1/4)^{1/2}) \qquad (11.3)$$
$$2^5 = \exp(\pi/(\lambda_2 - 1/4)^{1/2}) \qquad (11.4)$$

for the up-type and down-type sequences respectively. Then

$$[(\ln(32\pi)/\pi]^{-2} + 1/4 = = \lambda_1 = 0.714 \qquad \approx g_2 = 0.63 \qquad (11.5)$$

$$[(\ln(32)/\pi]^{-2} + 1/4 = = \lambda_2 = 1.07 \qquad \approx g_3 = 1.21 \qquad (11.6)$$

where we adjoin Coupling Constant g values of comparable value from the coupling constant table (Fig. 11.1).

Note the values of λ_1 and λ_2 are related by:

$$1.07/.714 = 1.5 \qquad (11.7)$$

Note the sum of the values $\lambda_1 + \lambda_2 = 1.784$ and also $g_2 + g_3 = 1.84$. These values differ by 3.4% suggesting that the ElectroWeak SU(2) and Strong SU(3) interactions combine to generate each mass sequence through an interplay of interactions.

11.7.3 Separation of SU(2) and SU(3) Contributions to the Sequences' Powers

We view the values in the sequences as sums of SU(2) and SU(3) terms. We calculate the discrepancy x between the λ_i from

$$\lambda_1 + x = 1.21 \qquad (11.8)$$
$$\lambda_2 - x = 0.63$$

using the approximate equality: $\lambda_1 + \lambda_2 = 1.784$ while $g_2 + g_3 = 1.84$. We find

$$x = 0.468 \qquad (11.9)$$

After noting that $g_4 = 2.4$ is itself an extrapolation we will use the approximate values from Fig. 11.1 to obtain

$$x = \tfrac{1}{2}(g_1 + g_2) = 0.466 \qquad (11.10)$$

Then we find an approximate relation between the SU(1), SU(2) and SU(3) coupling constants, and the λ_i, factors in in the fundamental fermion mass sequences.

$$\lambda_1 = 0.714 = 1.21 - 0.468 = g_3 - \tfrac{1}{2}(g_1 + g_2) \qquad (11.11)$$

for sequence 1, and

$$\lambda_2 = 1.07 = 0.63 + 0.468 = 3g_2/2 + \tfrac{1}{2} g_1 \qquad (11.12)$$

for sequence 2.

Sequence 2 is governed by the ElectroWeak interactions. Sequence 1 is governed by a combination of ElectroWeak and Strong SU(3) interactions. These results are reflected in the data since sequence 1 masses grow much more rapidly than sequence 2 masses.

The reasonableness of the eqs. 11.11 and 11.12 expressions for the multipliers of the sequences supports the choice of the sequence 1 multiplier $2^5\pi$ based on the t/c mass ratio, and the sequence 2 multiplier 2^5 approximately based on sequence 2 masses. They approximate the values in Fig. 11.3. The dependence of the multipliers of interactions would appear to be directly useful for specifying the dynamics of gambols in the author's Gambol Model presented in previous books. See the next section.

It is important to view the generation of fermion masses with Strong and ElectroWeak interactions as occurring *within* each individual fermion. The Higgs mechanism is used conventionally to generate vector boson masses that may play a role within interaction effects in fermions.

However our success in describing the fermion masses raises the question whether vector boson masses may also be similarly described. In addition, the masses of scalar bosons may also be so described. We have shown numerical relations between vector and scalar boson masses in Blaha (2024g) that are suggestive of this possibility. We will return to this issue in section 7.1, and in a future work.

This chapter directly relates fermion mass sequences to the coupling constant sequence.

11.8 Gambol Fermion Masses Related to ElectroWeak and Strong Coupling Constants

Since we have said the gambol fermion mass corresponding to a fundamental fermion is 1/844242 of the fermion's mass, the gambol mass must be the result of the same process that generates the fermion's mass. Therefore we apply the same logic as section 11.7 to relate the gambol mass to the SU(3), SU(2) and U(1) coupling constants: 1.21, 0.63, and 0.303 respectively. From Fig. 11.1 we see

U(1)	$\alpha_1 = \alpha$	$e^2/1024$	0.00721438	0.303	0.0072973525643	1.15%
SU(2)	$\alpha_2 = g^2/4\pi$	$e^2/256$	0.0289	0.63	0.0316	9.3%
SU(3)	$\alpha_3 = \alpha_S$	$e^2/64$	0.115	1.21	0.117	1.7%

For the multiplier 95939 the λ value is

$$\lambda = 0.325 \tag{11.13}$$

based on row E of Fig. 10.1:

E	n_φ Values Sequence; A Sequence of Sequences	½	95939	0.325

We have found the λ value for the gambol multiplier 95939 almost equal to the U(1) coupling constant:

$$\lambda = 0.325 = 0.303 + 0.025 \tag{11.14}$$
$$95939 = \exp(\pi/(\lambda - 1/4)^{1/2})$$

using the Sequence Theory equations:[36]

$$m = 16/\lambda_{n\varphi=4} = 0.7696 \tag{3.11}$$
$$\lambda_{n\varphi} = n_\varphi^{\,2}/m \tag{3.8a}$$
$$g_{n\varphi} = \exp(\pi/(\lambda_{n\varphi} - 1/4)^{1/2}) \tag{3.8b}$$
$$\lambda_{n\varphi} = [(\ln(g_{n\varphi})/\pi]^{-2} + 1/4 \tag{3.8c}$$

Thus we ***might** have* tentatively attributed the multiplier 95939 of gamboling primarily to U(1). The λ value in eq. 1.13 specifies the coupling constant for the $1/r^2$ interaction in the Master Equation for the row E sequence of masses. The choice of the gambol multiplier of 95939 would cause the fermion gambol of stratum n to have the same mass as the fermion of stratum n – 1. This coincidence would indicate a form of

---

[36] From Blaha (2025b). Note eqs. 3.8a and 3.8c above are only approximately in agreement.

degeneracy that would raise questions: Is the gambol fermion of stratum n the same fermion as the fermion of stratum n – 1? If not, does not the confinement of gambols and the freedom of fermions raise serious issues of the nature of the differentiating factor between them? If both have the same mass then what prevents transitions between them that ruins the confinement of gambols?

***Another possibility presents itself. If we assume the value λ = 0.303 is the correct value, then the gambol multiplier becomes 844242.* See line F in Fig. 11.5.**

Sequence	n_φ	g_g multiplier factor	λ_{n_φ}
?	$7.96 \approx 8$	$\sqrt{2}$	82.42
?	6	1.58	46.778
A1 Cosmos spaces dimension array column lengths $2^{r/2+2}$	4	2	20.79
A2 Cosmos unification space dimension r	4	2	20.79
B Coupling Constants Multiplier	2	4	5.39
C Fermion Sequence 2 Masses	1	$2/3 \times 2^5$	1.3
D Fermion Sequence 1 Masses	$½^{½}$	$3/2 \times \pi 2^5$	0.65
E n_φ Values Sequence; A Sequence of Sequences	½	95939	0.325
F New n_φ Values Sequence based on α_{QED}	$0.48 \approx 2^{½}/3$	844242	0.303
G The sequence of Unification Spaces	4	2	20.79

Figure 11.5. Extended Cosmos Theory Sequences with a new sequence determined by setting λ = .303, which is the value of the QED couping constant e which determines the Fine Structure Constant α. See Fig. 11.1. Line G is abstracted from Fig. 8.13 for unification spaces. The first two rows are inserted for completeness. The first row is for the case of a multiplier √2. Its n_φ value is an extension of the series of n_φ values 2, 4, 8 of rows A1 through B. The second and third rows are for the cases $n_\varphi = 6$ and $n_\varphi = 8$ which also may be of interest. The sequences associated with these rows are not known at present.

In this case the stratum's 1 and 2 mass spectrums are now discrete and non-degenerate. For example,

With Stratum 2 fermion M_2
Stratum 2 Gambol $M_2/844242$ much lower than stratum 1 fermion
Stratum 1 fermion $M_2/95939$
Stratum 1 Gambol $M_2/(95939 \times 844242)$

Thus a stratum 2 gambol is different from, and cannot decay to, a stratum 1 fermion.

Another example is the mass spectrum of a fermion with mass M of the 4th stratum is Fig. 11.6.

Stratum 4 fermion of mass M
Stratum 3 fermion M/95939
Stratum 2 fermion $M/95939^2$
Stratum 1 fermion $M/95939^3$

...

Stratum 4 Gambol $M_2/844242$
Stratum 3 Gambol M/(95939×844242)
Stratum 2 Gambol $M/(95939^2\times844242$
Stratum 1 Gambol $M_2/(95939^3\times844242)$

Figure 11.6. The sequence of masses for one fundamental fermion of mass M. The masses of the strata of fermions is much larger then the masses of the sequence of gambols.

11.8.1 Fermion Mass Spectrum for Strata and Gambols

Now there are two multipliers appearing in the mass spectrum. The 95939 multiplier is used to separate the fermions' masses by stratum. The 844242 multiplier is a "fine structure" factor separating each fermion from its corresponding gambol fermion.

The masses of the fermions are apparently set by SU(3) and SU(2) interactions. The gambols appear to be generated by a U(1) QED-like interaction.

It is important to realize that each r = 4 UST layer has independent, differing SU(3), SU(2) and U(1) gauge fields and interactions in the Normal and Dark sectors. For r > 4 universes there are $2^{r-1} = 8$ sets of ElectroWeak-like and Strong gauge fields and interactions – each acting on the fermions of its subsector.

11.8.2 Origin of Gambols in QED

The role of U(1) in determining the 844242 multiplier suggests that each pair of fundamental charged fermions of the 256 fermions in the UST's fermion dimension array is split by the electromagnetic interaction into a state consisting a combination of the fermion and a gambol fermion partner:

$$\begin{bmatrix} \psi_{\text{Fermion}} \\ \psi_{\text{Gambol}} \end{bmatrix} \tag{11.15}$$

Since the electromagnetism interaction presumes a charged fundamental particle the splitting of the fundamental fermions of UST applies only to charged fermions. We thus envision the masses of the charged fermions due to the ElectroWeak and Strong interactions and the splitting of each of the charged fermions into a fundamental fermion and a gambol fermion due to a QED-like interaction.

Since neutrinos have no charge they do not have gambol fermion partners. They do occur in the generations and layers of Normal and Dark fermions. And they do appear in quadruplicate in the four strata.

11.8.3 Implications for UST Part of Cosmos Theory

The QED basis of the split fermion-gambol fermion pairs has numerous implications. First we note that there are four QED's in the UST Normal sector. See Fig. 8.2. Each of the UST's four fermion layers have a different SU(2)⊗U(1) containing QED. The fermion-gambol splitting applies to each UST layer in each stratum based on its QED.

The total number of UST fermions (including gambols) is based on:

256 fermions of which 224 are charged
32 Normal and Dark neutrinos

The total number of UST fermions (including gambols) is:

$$224 \times 2 \text{ states} \times 4 \text{ strata} + 32 \times 4 \text{ strata} = 1920 \text{ fermions} \qquad (11.16)$$

11.8.4 Implications for Dimension r Cosmos Theory Space

The total number of dimension r fermions (including gambols) is based on:

2^{r+4} fermions of which $2^{r+4} - 2^{r+1}$ are charged
2^{r+1} Normal and Dark neutrinos
$2^{r+4} - 2^{r+1}$ possible fermion and gambol pairs

The total number of dimension r universe fermions including gambols is:

$$(2^{r+4} - 2^{r+1}) \times 2 \text{ states} \times 4 \text{ strata} + 2^{r+1} \times 4 \text{ strata} = 2^{r+7} - 2^{r+4} + 2^{r+3} \text{ fermions} \qquad (11.17)$$

Note that the size of the dimension array is still 2^{r+4}. The total number of dimension r fermions counting strata but not counting gambols is 2^{r+6}. The total number of dimension r fermions counting strata and gambols is $2^{r+7} - 2^{r+4} + 2^{r+3}$.

11.9 Sequence Theory in the r = 4 UST

Some of the above results were based on our r = 4 UST. We now show the Sequence Theory results also apply to other spaces and universes of higher dimension.

11.9.1 Extension of UST Results to Other Generations and Layers

The results in Row B for Coupling Constant sequence apply only to the First layer of the UST. Each of the other three layers have a different set of Internal Symmetry groups for SU(4) (or SU(3)⊗U(1)), SU(2)⊗U(1), Generation U(4), and Layer U(4). We assume the coupling constants in these layers have the same multiplier of 4 but differ in their overall constant labeled α_0 in Fig. 11.1. Our derivation of the full dimension array from an FRF supports the consistency of the coupling constants behavior.

Similarly, the fermion mass sequences in Rows C and D for eight fermions were for the first generation of the first layer. The other 31 corresponding sets of 8 fermions in Normal and Dark sectors can also have similar sequences up to overall constants. This possibility remains to be confirmed experimentally.

The row E sequence, which leads to strata, is extended to the other generations and layers directly.

Thus these sequences may be extended throughout the UST.

11.10 Extension of UST Results to Other Spaces and Universes

The sequence results of the types of rows A1 and A2 apply directly to Cosmos Theory spaces. The sequences corresponding to rows B, C and D generalize directly to higher dimension spaces and universes due to the quadrupling feature of dimension arrays and their associated fermion and internal symmetry structure.

The fermions of higher dimension r spaces and universes number $2^{r+7} - 2^{r+4} + 2^{r+3}$ based on four strata of fermions and a gambol fermion for each charged fermion of the space.

The sets of 8 fermions that form the two sequences described above are replicated 2^{r+4} times in the space and universes of dimension r. This replication is consistent with the reduction of dimension arrays by GRIS transformations to an FRF vector with $2^{r/2+2}$ non-zero components. The non-zero components total to $2^{r/2-1}$ sets of 8 fermions. Each set can be assumed to have two mass sequences similar to those seen above but with different overall constants.

12. The Basis of Universes

We have discussed aspects of universes in detail in previous books. In this chapter we overview some important points of universes created in Cosmos Theory.

Earlier we had discussed the Consistency Condition that presides over the specification of the number of HyperCosmos spaces limiting the number of spaces to 10. This condition also specifies a Parent universe within which all subuniverses reside. (There can be more than one Parent universe.) The structure of the Cosmos spaces exists forever in the sense that there is no time. The Parent universe also exists forever. There is no beginning and no end.

Subuniverses are created, directly, or indirectly, within the Parent universe. See Fig. 12.1. Subuniverses are universes whose spaces are one of the Cosmos spaces.

There are a number of possible mechanisms for the creation of subuniverses. One mechanism is to treat subuniverses as "particles" and to posit that they are the entities in a Lagrangian quantum field theory of universes. Then they could be created from the Parent "vacuum" perhaps or as a result of universe interactions. We have considered this possibility in earlier books.

12.1 Their Nature at Creation

If subunivereses are generated in a Quantum manner, then it is reasonable to believe they are point-like surrounded perhaps with vacuum polarization. We expect that subuniverses contain enormous energy (Big Bang) and are likely to expand.

The initial state of a subuniverse is pure energy embodied in the motions of the constituents of the subuniverse. The particles (waves) in the subuniverse must be initially massless due to the extreme high energy of the subuniverse. *All subuniverse mass-energy is initially energy of motion of massless particles.*

12.2 Hierarchies of Universes

Cosmos Theory supports a Hierarchy of Universes. There is a Parent universe that is coeval with the set of Cosmos spaces. We developed that concept in our Consistency Condition for the Cosmos. Then subuniverses can be created within the Parent universe through some mechanism. (We discussed possible mechanisms in earlier books.) The creation of subuniverses can be nested as we see in Fig. 12.1.

12.3 Particle Models of Universes

We have created Quantum Field Theory particle models of universes with interactions in previous books to which the reader is referred. (See the References Section and particularly Blaha (2019c) and (2019d).) We have extended these models to include Gambol Models for Hubble expansion of universes, S_8 and other universe features.

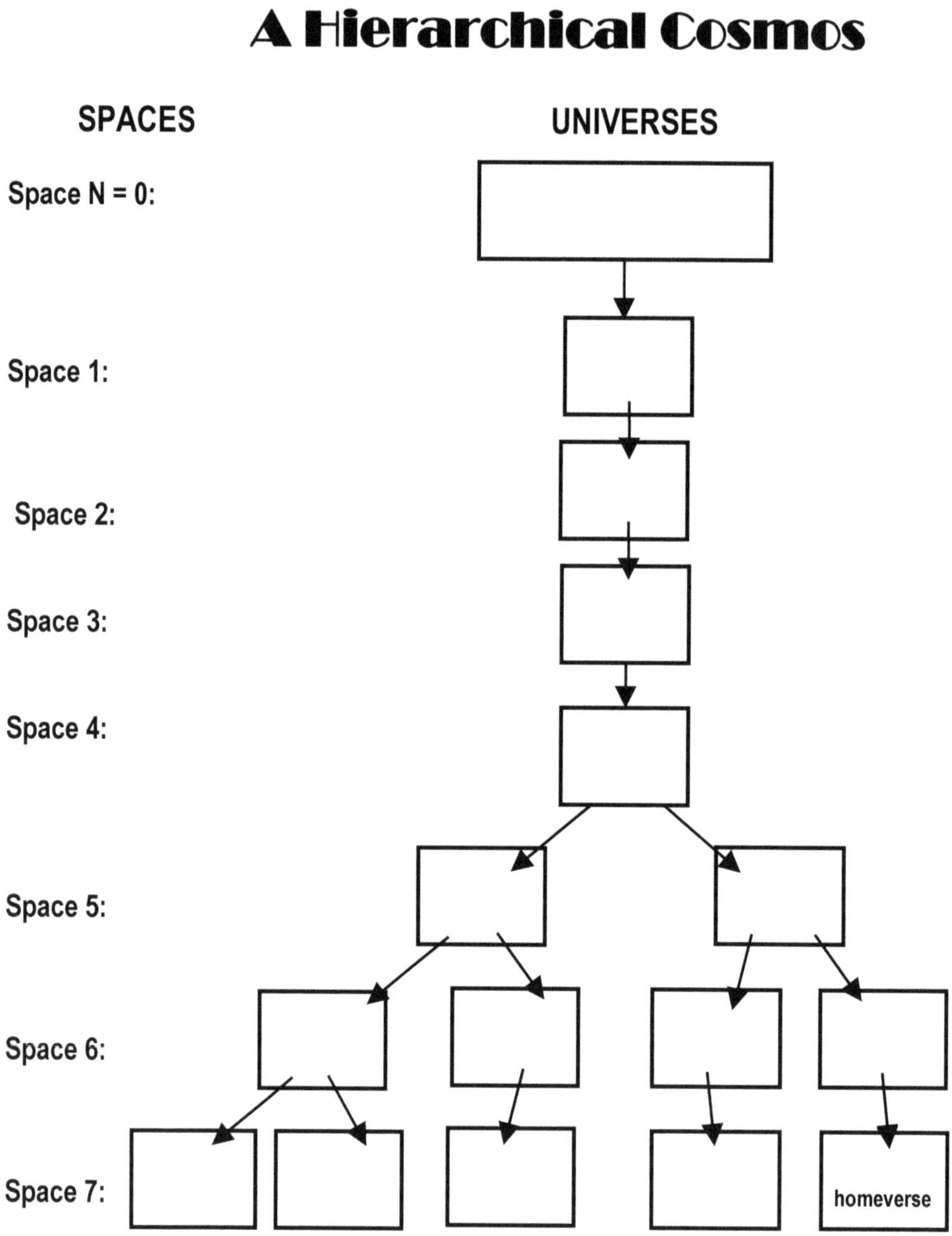

Figure 12.1. A hierarchy of universes (and possibly anti-universes) leading from the Parent universe (N = 0) to subuniverses.

13. What is Mass-Energy?

In chapter 12 we briefly described the creation of a subuniverse. All constituents of the universe at the creation point were initially massless due to the extremely high temperature. Thus all energy was "energy of motion." The energy of motion was embodied in the fermions, gambol fermions, and bosons within the universe – if there were any. (It is possible that the initial state of the universe was simply a state of vacuum energy.)

As a subuniverse expands its temperature decreases. Eventually particles appear. These particles acquire masses. These particles are fermions, gambols, and bosons. We have found that fermions have constituents (gambols). The constituents are in motion. They have nominal masses that are 1/844242 of the masses of their fermion partners. Vector bosons also acquire masses through the Higgs Mechanism.

The consideration of the nature of masses thus goes down to a much more refined level. As a subuniverse temperature increases from its initial value masses acquire value in a consistent manner as evidenced by the mass sequences and coupling constants that we found.

If we were to inquire as to the nature of masses beyond the fact that they appear as constants in Lagrangians and fundamental equations, we would have to say that masses are simply embodiments of energy that are made into "unchanging" constants.

We are thus led to view all energy, whether directly in the form of motion, or indirectly within bodies disguised as particles with masses, as an embodiement of motion.

14. The Fundamental Element

The progress of Physics has gone from four elements to one element to atoms to fundamental particles to gambols. We have suggested that there is a new form of element.

Dimension is the basic element.

The proof of this assertion may be seen in the earlier chapters of this book (and in our earlier books) where there is a progression from a point (actually a dimension) through dimensions up to two dimension unit squares – through the Master Equation to a kaleidoscope of dimensions embodied in spaces – to universes of many dimensions.

15. Unified SuperStandard Theory (UST) with Strata and Gambols

We developed the Unified SuperStandard Theory (UST) as an extension of the Standard Model of Elementary Particles in the period from 6 to 11 years ago. Six years ago it became apparent that the UST could naturally be viewed as a type of theory based on Cayley Numbers (quaternions, octonions, and so on). As a result we have spent the past 6 years developing Cosmos Theory, Sequence Theory, and Gambol Theory

In this chapter we present a succinct description of major features of the UST. (Some new UST features were described in earlier chapters.) We add more detail on the implications of the new fermion strata and gambols. The enhanced UST now stands as a distinct theory awaiting further experimental confirmation.

15.1 UST Dimension Array

For r = 4 the UST dimension array has $2^{r+4} = 256$ components (dimensions). It is depicted in Fig. 15.1 using dots to represent dimensions. The dimension array specifies the pattern of 256 fermions in the UST in Fig. 15.2. This figure treats the fermions as if they are of the same stratum. Later we will discuss the strata and suggest a new set of fermion patterns.

15.2 Fermion Spectrum

The fermions in Fig. 15,2 were distributed in a direct manner:

1. The array was separated into Normal and Dark sectors
2. Each row contained a generation of fermions.
3. Since 3 generations were known and one more generation seemed likely, we separated the array vertically into four sets of four generations. We called each set of 4 generations a layer.
4. In the first row we placed 8 of the known fermions. The remaining μ and τ neutrinos were placed in the 2nd and 3rd rows respectively as 2nd and 3rd generation fermions.
5. The Dark sector was assumed to mirror the Normal sector with different particles.

15.3 Symmetry Groups Spectrum

The dimension array was also mapped to the known Standard Model symmetry groups with the addition of Generation groups and Layer groups. The symmetry groups of the UST were distributed in a direct manner as well. See Fig. 15.3.

The symmetry groups were also distributed in a transparent manner:

1. They were distributed by layers.

2. They were distributed as similar, but different, Normal and Dark sector groups.
3. The first layer had the known Standard Model groups SU(3)⊗U(1) SU(2)⊗U(1)⊗SL(2, C) plus U(4) Generation groups and U(4) Layer groups. Appendix 15-A describes the role of the Generation and Layer groups. These groups were necessary for interactions between generations, and interactions between corresponding fermions in different layers as shown in Fig. 15.2.
4. Each of the four layers had their own set of groups as shown in Fig. 15.3. Thus there are 32 different groups distributed across four layers.
5. Within a layer we separated the groups into parts, with each part using 8 real dimensions. See Fig. 15.3.
6. Each Layer group representations had fermions in each of the four layers. For convenience, we placed a Layer group in each layer although their representations overlapped all 4 layers.

Thus the above distribution specification is realized in Fig. 15.3.

It is somewhat remarkable to see how well adapted the array of symmetry groups corresponds to the array of fermions, for which they provide interactions. We will discuss the UST interactions in conjunction with strata and gambols later.

The Generation and Layer groups were originated by this author many years[37] ago when he saw the form of the generations and layers of fermions in the 256 dimension array of the UST. They were not conceived in a "vacuum."

For each layer, the Generation group provides interactions between the fermions of the four generations of fermions of the layer. Each of the four layers have a separate Generation group for its generations. The Generation groups implement interactions between fermions of different generations. These interactions are necessarily weak. See Appendix 15-A for details.

The four Layer groups provides interactions between the fermions of different layers. They unite the framework of the four layers of fermions. For example, one Layer group has representations of four corresponding fermions consisting of a fermion from each layer. Each of the corresponding fermions have the same internal symmetry quantum number values and generation number. Therefore the only quantum number difference is the layer number. The Layer groups implement interactions between fermions in different layers. These interactions are necessarily very weak. Appendix 15-A has the details.

15.4 Connection Groups

The Connection groups are SU(2) groups that provide interactions in two ways. First, they provide interactions between corresponding Normal and Dark fermions. We call these interactions horizontal interactions. They also provide interactions vertically between corresponding fermions in adjacent layers. These interactions are not Layer

---

[37] See Blaha (2018e) for a description of Generation and Layer groups. These groups originally appeared in earlier books by the author.

group interactions because they are specific to pairs of layers. Layer group interactions specify interactions among fermions in all four layers. See Fig. 15-A.2.

With the introduction of gambols the Layer groups must be extended to perform similar transformations, with similar interactions, for gambol fermions as well as fermions. Fig. 15-A.2 illustrates that point.

15.5 Strata Extension of the Fermion Spectrum

The fundamental fermion spectrum of Fig. 15.2 must be revised when fermion strata are introduced. We replace the spectrum with a new spectrum consisting of four instances of the fermion array of Fig. 15.1. The result appears in Fig. 15.4 together with gambol fermion strata. Each fermion of each stratum has a corresponding fermion in the other three strata. Each fermion has four multiplicates. The only difference, other than stratum number, is their masses. The masses increase from the lowest stratum masses by factors of 95939 – stratum by stratum – as we described earlier.

We have placed the known neutrinos at the first stratum. We have placed the known charged fermions on the 2nd stratum for reasons described in earlier chapters.

15.6 Symmetry Breaking Through Fermion-Gauge Boson Interactions

The introduction of Strata and Gambols, and symmetry groups with interactions, into the UST leads to questions of symmetry breaking by the existing and new interactions. The general form of the interactions is

$$g\, \bar{\psi}_k A^{a\mu} \gamma_\mu T_a \psi_j \tag{15.1}$$

where j and k are field operator labels, $A^{a\mu}$ is a gauge field of a symmetry group, and the set of T_a are the group generator matrices.

The symmetry groups have the following forms for the various parts of the complete spectrum of fermions and gambol fermions:

1. The groups: SU(4) or SU(3)⊗U(1); SU(2)⊗U(1); and Generation U(4) group are the same for all generations in a layer. They differ from layer to layer.
2. The Layer groups "straddle" the layers with fermions in representations spread across all four layers. The fermions in a representation have the same quantum numbers except for the layer number, which ranges from 1 through 4. An interaction of the form of eq. 15.1, where j and k are layer numbers and all other quantum numbers are the same, has a field operator which causes a field operator with j type internal symmetries to become a field operator with k type internal symmetries. As a result there is symmetry breaking for both the j type symmetries and the k type symmetries induced by the interaction. This type of symmetry breaking must be ultra weak – much weaker than ElectroWeak symmetry breaking.
3. The U(2) Connection Groups also have interactions that lead to symmetry breaking. The vertical Connection Groups (E, F, G) in Fig. 15-A.2 have the same sort of symmetry breaking as described above in item 2. The difference

between the Layer and Connection groups in this case is in the U(4) of Layer Groups and the U(2) of Connection Groups. The vertical Connection groups have representation of two corresponding fermions where each can be in different layers.

The horizontal Connection Groups (A, B, C, D) of Fig. 15-A.2 relate Normal fermions with corresponding Dark fermions. Assuming the groups have equal forms for both Normal and Dark sectors – but are different, an interaction of the type of eq. 15.1 couples a Dark fermion with a corresponding Normal fermion causing symmetry breaking of both Normal and Dark symmetries. These types of symmetry breaking must be extremely small since we have not seen Dark fermions as yet.

4. The introduction of Strata leads to the possibility of 6 Strata Symmetry Group interactions. See Fig. 15.5. These groups would have interactions of the form of eq. 15.1 between corresponding fermions in different strata. Since the strata have identically the same symmetry groups, the only symmetry breaking would be in the stratum number. (We give a stratum number to each fermion in each stratum. We have previously assigned generation numbers and layer numbers to fermions in the UST.)
5. The introduction of strata implies the Connection Groups must be generalized to connect fermions. We stipulate the strata are "diagonally" connected for each stratum. So the Connection group representations are quadrupled.
6. The introduction of gambols implies the Connection Groups must be generalized to connect gambol fermions. We require the gambols are "diagonally" connected.
7. The Gambol Strata Groups can have an interaction of the form of eq. 15.1 between corresponding gambol quantum fields of different strata. See Fig. 15.6. These interactions would break the stratum number for gambols but would not break the quantum numbers of other symmetry groups.

15.7 Strata and UST Symmetry Groups

The symmetry groups of the UST are extended when fermion strata are introduced. The form of the symmetry groups extension is:

1. The same symmetry groups apply to all strata. The quantum numbers of all corresponding fermions are identical except for stratum number. The stratum number is a new quantum number. Its value for each fermion is the fermion's stratum number.
2. Interactions between fermions of the symmetry groups are "blind" to the stratum number. The internal symmetry interaction terms for fermions in the full Lagrangian are independent of stratum number.
3. The symmetry groups and Higgs bosons are independent. They are not "multiplicated."

4. There is a possibility that the sequence of fermion masses based on factors of 95939 are due to a Higgs mechanism. This prospect remains to be explored.
5. It is possible to define six[38] interaction symmetry groups for strata that support transitions between corresponding fermions in different strata. In the absence of a differentiating factor for interactions between fermions of symmetry groups described in sections 15.3 and 15.4 we will not pursue this possibility here.

15.8 Strata Symmetry Groups

We can define new symmetry groups, *strata symmetry groups*, that create interactions between fermions of different strata. The interaction between corresponding fermions of stratum j and stratum k has the form

$$\bar{\psi}_k A^{a\mu} \gamma_\mu T_a \psi_j \tag{15.1}$$

where $A^{a\mu}$ is a gauge field of a strata symmetry group, and all quantum numbers are the same except for the stratum number.

There are 6 strata symmetry groups. Their roles are shown in Fig. 15.5. Whether these groups are relevant depends on whether differentiating features and/or interactions are introduced between the fermions of the various strata.

15.9 Strata Gambol Fermions

Each stratum of fundamental fermions has a subsidiary stratum of gambol fermions.

The 1/95939 factor between fermion strata masses is due to the $n_\varphi = ½$ sequence which appears in Fig. 11.5. Fig. 11.5 also shows a new $n_\varphi = 2^{½}/3$ sequence with multiplier 844242 that is due to using $\lambda = e$, the QED coupling constant.[39]

For each fermion in each stratum the mass of the corresponding gambol fermion is 1/844242 of the fermion mass. The use of the 1/844242 factor for relating gambol masses to their fermion's mass (rather than 1/95939) prevents mass degeneracies. As a result the spectrum of fermion strata masses is disjoint from the spectrum of gambol fermion masses. See Fig. 15.4.

15.10 New Gambol Strata Groups

The four strata of gambol fermions leads to the consideration of the possibility of *Gambol Strata interaction groups*. The Gambol Strata groups support interactions between corresponding gambols of different strata. Fig. 15.6 displays the six interaction groups for four strata of gambols. The form of the six Gambol Strata groups parallels the form of Strata Symmetry groups discussed in section 15.7.

15.11 The Groups of the UST, Strata, Gambols

The introduction of Strata and Gambols into the UST introduces more possible symmetry groups and their interactions. The currently complete list of groups is:

---

[38] They would be analogous to the gambol-strata interaction groups.

[39] In chapter 11 we showed a close relationship between coupling constants and λ values for fermion mass sequences. Compare line F in Fig. 11.5 and the U(1) QED coupling constant in Fig. 11.1.

1. The UST symmetry groups.
   SU(4) or SU(3)⊗U(1); SU(2)⊗U(1); U(2) Connection groups; Generation U(4) groups; Layer U(4) Groups.

2. The Strata Groups
   6 Strata Symmetry groups - Section 15.8.

3. The Gambol Strata Groups
   6 Gambol Strata Symmetry Groups - Section 15.10 and Fig. 15.5.

4. Additional Connection Groups
   The 7 U(2) Connection Groups which connected fermions are now expanded to include connecting corresponding gambols.
   The introduction of strata implies the Connection Groups must be generalized to connect fermions of different strata. We stipulate the strata are "diagonally" connected for each stratum. The Connection group representations are quadrupled. See Fig. 15-A.2 and Fig. 15.6.

Figure 15.1. The 128 dimensions **half** of the r = 4 HyperCosmos dimension array of 256 dimensions. The dimension array is square with the rows and columns: 16×16. We assign half of the 256 dimensions to the Normal matter sector and half to the Dark sector. The list of fermions at the top of the figure are Standard Model fermions that correspond to each filled top circles when the dimension array is mapped to fermions. Each fermion characterizes the fermions in the column beneath it (subject to some quantum number differences).

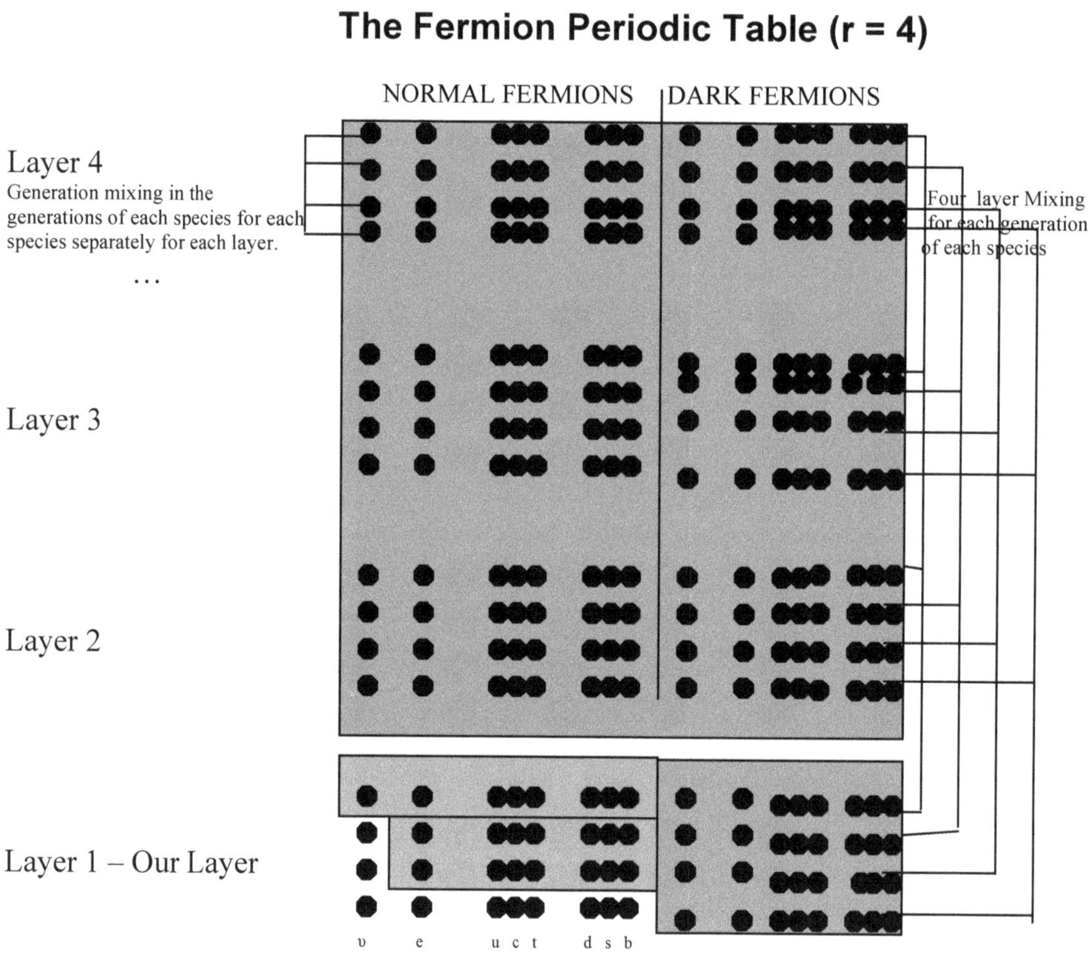

Figure 15.2. The r = 4 UST Fermion particle spectrum and partial examples of the pattern of mass mixing of the Generation groups and of the Layer groups. Unshaded parts are the known fermions, A shaded generation is an additional, as yet not found, 4th generation. The lines on the left side (only shown for one layer) display the Generation Group mixing within each layer. The Generation mixing occurs within each layer using a separate Generation group for each layer. The lines on the right side show Layer group mixing (for Dark matter) with the mixing among all four layers for each of the four generations individually. There are four Layer groups for Normal matter and four Layer groups for Dark matter. There are 256 fundamental fermions. The fermion types are labeled on the low line of the display.

Layer	NORMAL		DARK	
1	SU(3)⊗U(1) Generation U(4)	SU(2)⊗U(1)⊗SL(2, C) Layer U(4)	SU(3)⊗U(1) Generation U(4)	SU(2)⊗U(1)⊗U(2) Layer U(4)
2	SU(3)⊗U(1) Generation U(4)	SU(2)⊗U(1)⊗U(2) Layer U(4)	SU(3)⊗U(1) Generation U(4)	SU(2)⊗U(1)⊗U(2) Layer U(4)
3	SU(3)⊗U(1) Generation U(4)	SU(2)⊗U(1)⊗U(2) Layer U(4)	SU(3)⊗U(1) Generation U(4)	SU(2)⊗U(1)⊗U(2) Layer U(4)
4	SU(3)⊗U(1) Generation U(4)	SU(2)⊗U(1)⊗U(2) Layer U(4)	SU(3)⊗U(1) Generation U(4)	SU(2)⊗U(1)⊗U(2) Layer U(4)

Figure 15.3. The symmetries of the UST. The undarkened row is for the known symmetry groups. The darkened parts have not as yet been found. Note each group item (There are 32 items.) has an 8 real dimension representation. The SL(2, **C**) representation has four real coordinates and thus occupies 4 dimensions.[40]

[40] The Lorentz Group $SO^+(1, 3)$ is often specified with an SL(2, **C**) representation.

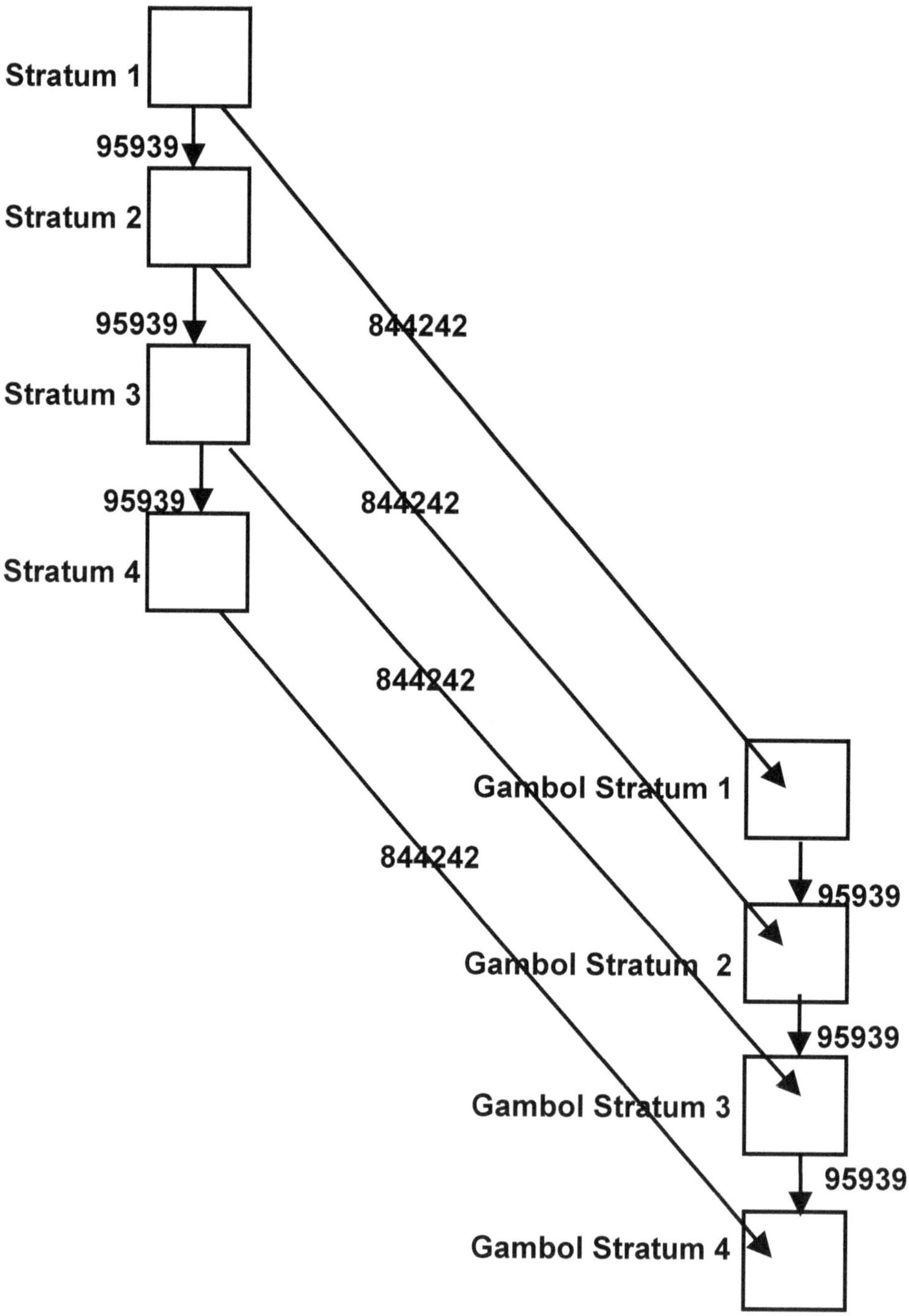

Figure 15.4. Each left block represents a 256 fermion set. Each right block represents a corresponding 224 gambol fermion set since neutrinos do not have gambols. Coresponding fermions in adjacent strata differ in mass by a factor of 95939 in increasing order from stratum 1. Each gambol fermion is a factor of 1/844224 of the corresponding fermion mass. Each gambol mass is 1/95939 of the mass of the gambol immediately above it. Not drawn to scale.

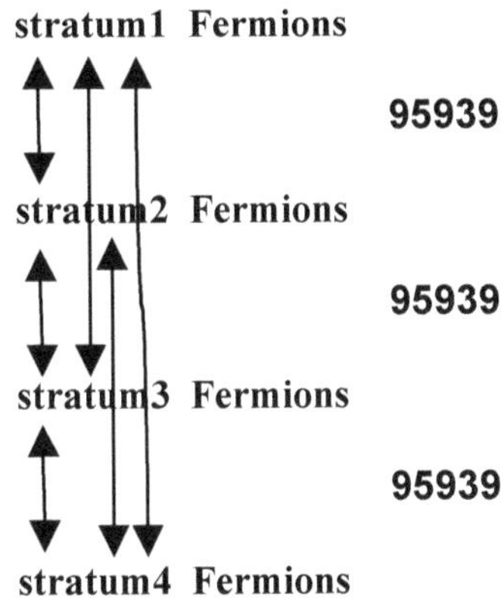

Figure 15.5. A view of the four UST strata of the Sequence Theory showing the 6 possible groups and interaction roles relating the fermions in the various strata. The interaction indicated with the arrows is between a fermion of one stratum with the corresponding fermion of another stratum. For each fermion type the masses increase by a factor of 95939 as one descends the strata. The additive quantum numbers are the same as one descends the strata.

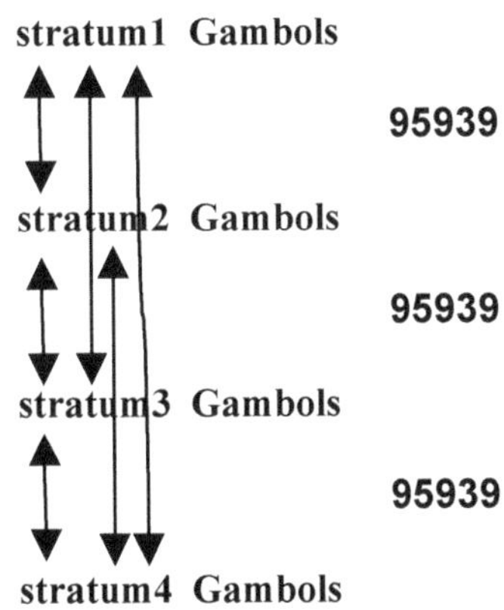

Figure 15.6. A view of the four Gambol strata of the Sequence Theory showing the 6 groups and interactions relating the gambols in the various strata. The interaction indicated with the arrows is between a gambol of one stratum with the corresponding gambol of another stratum. For each gambol type the masses increase by a factor of 95939 as one descends the strata due to the increase of fermion masses. The gambol additive quantum numbers are the same as one descends the strata. Gambol masses are a factor of 1/844242 of the fermion masses in the same strata. In all strata the gambol additive quantum numbers are 1/844242 of the fermions' additive quantum numbers. Thus the gambols' additive quantum numbers of **all** strata can be taken to be 1/844242 of the additive quantum numbers of the fermions of stratum 1.

Appendix 15-A. Connection Group Symmetries and HyperCosmos Space-Time Coordinates

15-A.1 Hypercomplex Coordinates Transformed to Symmetry Groups in the UST

Our UST formulation for our r = 4 universe has a d_{d7} = 256 component dimension array. A preliminary view of the symmetries that it contains appears in Fig. 15-A.1.

Note that there are 32 dimensions (in the leftmost column) that are initially allocated to space-time dimensions suggesting a hypercomplex space-time. We chose to allocate 4 real dimensions to obtain the 4-dimension space-time implied by eq. 15-A.1. The remaining 28 real dimensions we allocated to additional symmetry groups—namely seven U(2) groups. We call these groups Connection groups since their role is to "connect" fermions residing in different fermion spectrum blocks.

The structure of the seven additional U(2) groups is not specified. We chose to use a reasonable physical principle to allocate them. We believe their role is to "connect" corresponding fermions in different blocks since the fermions within each fermion block have "known" interactions.

Note also that there are eight blocks of 32 dimensions in Fig. 15-A.2, each with their own set of symmetry groups and corresponding interactions, and initially no interactions between the eight blocks. If there are no block interactions (except gravity), then the Physics of fermion and symmetries sets are conceptually disjoint.

We choose to implement inter-block interactions as in Fig. 15-A.2 following the principle:

A fermion in any block has interactions either directly, or indirectly, with every other fermion in every other block.

Fig. 15-A.2 shows an implementation of this principle. The horizontal lines in Fig. 15-A.2 indicate 1:1 transformations between all corresponding fermions of each "Normal" and "Dark" block. The three "angled" lines indicate 1:1 transformations between corresponding fermions of a "Normal" and a "Dark" fermion block in the layer above it. The result is a see-saw type of pattern. We take each Connection group to be a U(2) group because of its representation of 4 real dimensions.

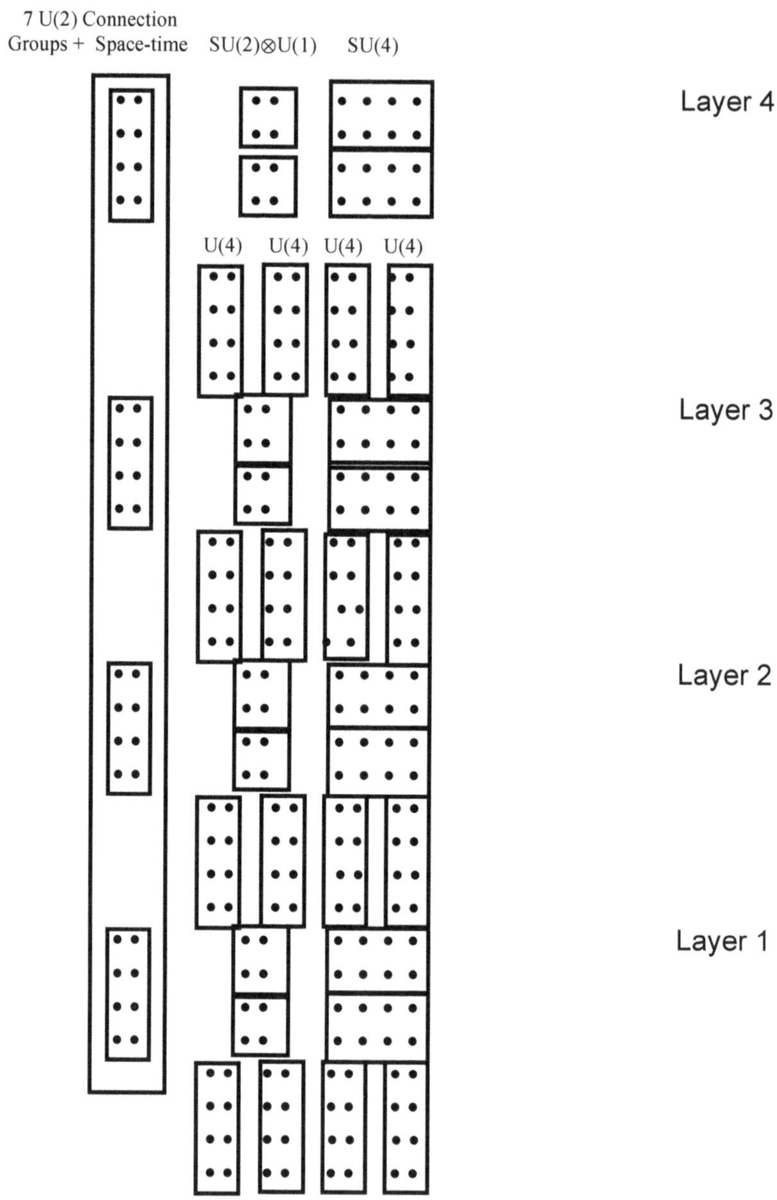

Figure 15-A.1. The four layers of UST symmetry groups (including space-time) including both Normal and Dark sectors. There is an SU(4) symmetry before breakdown to SU(3)⊗U(1). Note the left column of blocks are combined below to specify a 4 dimension real space-time plus seven U(2) Connection groups. Note each layer has 64 dimensions = 56 + 8 dimensions.

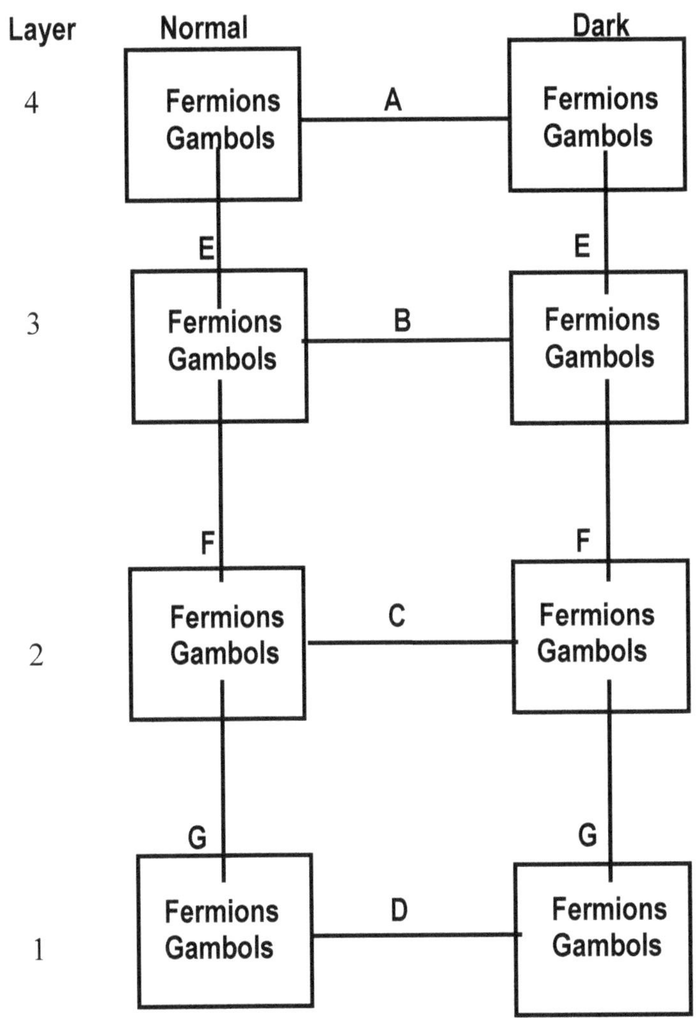

Figure 15-A.2. The seven U(2) Connection groups (shown as 10 lines) between the eight UST blocks. Connection groups are obtained by transfering 28 dimensions from UST space-time to internal symmetries with the consequent reduction of the space-time from four octonion (complex quaternion) coordinates to four real coordinates. The Connection groups generate rotations and interactions between corresponding fermions and vector bosons of each pair of blocks. The Normal and Dark sector U(2) vertical connections above (E, F, G) represent the same U(2) groups. The Connection groups perform the same chores for Gambol Fermions as shown.

15-A.1.1 The U(2) Connection Groups

The seven U(2) Connection groups of Fig. 15-A.2 generate "rotations" and interactions between *corresponding* fermions and vector bosons of each pair of blocks of the eight blocks of fermions in the UST.

15-A.1.1.1 Horizontal Lines

The horizontal lines in Fig. 15-A.2 (A, B, C, and D) each represent a U(2) Connection group that "rotates" two *corresponding* fermions in the Normal and Dark sectors of each layer. Thus a Normal e is "rotated" with a corresponding Dark e, and so on.

Each of the four horizontal Connection Groups has a reducible U(2) representation D that is the sum[41] of 4*8 = 32 irreducible U(2) representations. We may view the irreducible representations D_j as strung along the diagonal.

$$D = \sum_{j=1}^{32} D_j \tag{15-A.2}$$

for each of the U(2) groups of the four horizontal lines in Fig. 15-A.2.

The U(2) group also specifies gauge field interactions between corresponding fermions in each layer of the Normal and Dark sectors of the form

$$g\overline{\Psi}_{Nn}\gamma\cdot A\cdot T\Psi_{Dn} \tag{15-A.3}$$

where N indicates a Normal fermion and D indicates the corresponding Dark fermion, with A being a U(2) gauge vector boson, and n the label for corresponding fermions.

These U(2) transformations imply that the Normal and Dark sectors have the same species.

15-A.1.1.2 Vertical Lines

The pairs of vertical lines in Fig. 15-A.2 (E, F, G) each represent a U(2) Connection group that "rotates" sets of two *corresponding* fermions in adjacent layers as shown in Fig. 15-A.2 in the Normal and Dark sectors. Thus a Normal e in layer 1 is rotated with a corresponding Normal e in layer 2, and so on.

Each of the three (six counting both Normal and Dark lines in Fig. 15-A.2) vertical Connection Groups has a reducible U(2) representation D that is the sum of 64 irreducible U(2) representations.[42] We may view D as an array of 64 U(2) irreducible representation dimensions D_j strung along the diagonal.

$$D = \sum_{j=1}^{64} D_j \tag{15-A.4}$$

for each of the U(2) groups of the 3 (6) horizontal lines in Fig. 15-A.2. Note the 64 irreducible representations include both Normal and Dark sectors of a layer. [43]

Each U(2) group also specifies a gauge field interaction between corresponding fermions in adjacent layers for both Normal and Dark sectors:

---

[41] Eight fermions per generation ▪ four generations, thus accounting for each fermion in a block.

[42] Eight fermions per generations ▪ four generations ▪ 2 types of matter (Normal and Dark).

[43] There are 64 fermions in total for each of the four layers of the UST.

$$g\overline{\Psi}_{nl_1}\gamma\cdot A\cdot T\Psi_{nl_2} \qquad (15\text{-}A.5)$$

where l_1 and l_2 designate layers, A is a gauge field vector boson, and n the label for corresponding fermions.

Each E, F, and G U(2) group reducible representation includes both Normal and Dark sectors.

15-A.1.2 The Connection Groups are UltraWeak Interactions

Since there is no convincing experimental evidence for particle interactions between Normal and Dark matter, or between Normal fermion layers the Connection groups appear to be UltraWeak.

15-A.2 A Megaverse with Six Real Space-Time Coordinates (Dimensions)

This space is an r =6 HyperCosmos space. We view it as the Megaverse (Multiverse) with 6 real-valued space-time coordinates by eq.15-A.1 above. It has a d_{d6} = 1024 dimension array.

We view its dimension array as composed of four copies of the r =4 dimension array. Initially 4·32 = 128 dimensions are for space-time coordinates.

We allocate 4·28 = 112 dimensions to each of the four UST copies within the Megaverse to give each copy 7 SU(2) Connection groups. Therefore 4* 28 = 112 of the 128 space-time dimensions of the Megaverse are mapped. The remaining 16 dimensions give 6 real-valued space-time dimensions to the Megaverse space, 8 dimensions to a Megaverse SU(4) Connection group, and 2 dimensions to a U(1) group for all fermions in the Megaverse. See Figs. 15-A.3 and 15-A.4.

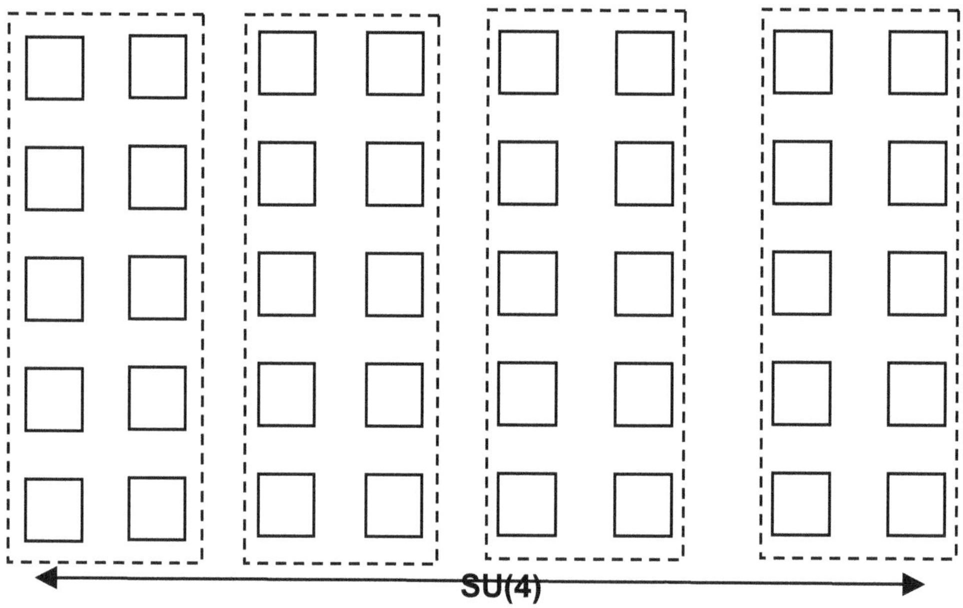

Figure 15-A.3. The Megaverse has four QUeST copies. An SU(4) internal symmetry Connection group maps between corresponding fermions in the four copies: fermion by fermion. An additional U(1) Connection group applies to every corresponding fermion. It is not shown in this figure.

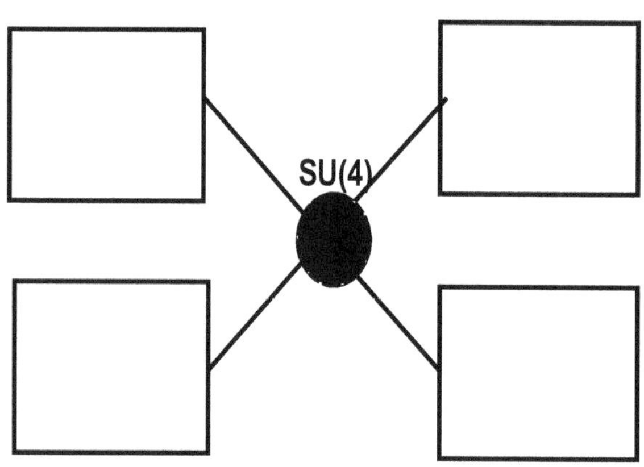

Figure 15-A.4. The SU(4) Connection Group of the Megaverse connecting fermions in the four UST "copies" blocks. An additional U(1) Connection group applies to every corresponding fermion. It is not shown in this figure.

15-A.3 A Maxiverse with Eight Real Coordinates (Dimensions)

The Maxiverse is the r = 8 HyperCosmos space with a dimension array of d_{d5} = 4096 dimensions.

The Maxiverse contains four copies of the Megaverse. Each Megaverse has 4*6 = 24 space-time dimensions. (The remainder in each Megaverse copy consists of internal symmetries and Connection groups.) We allocate the 24 dimensions to eight real space-time dimensions plus 16 dimensions for a new ultra weak (possibly broken) SU(8) Connection group for the four parts of the Maxiverse. See Fig. 15-A.5.

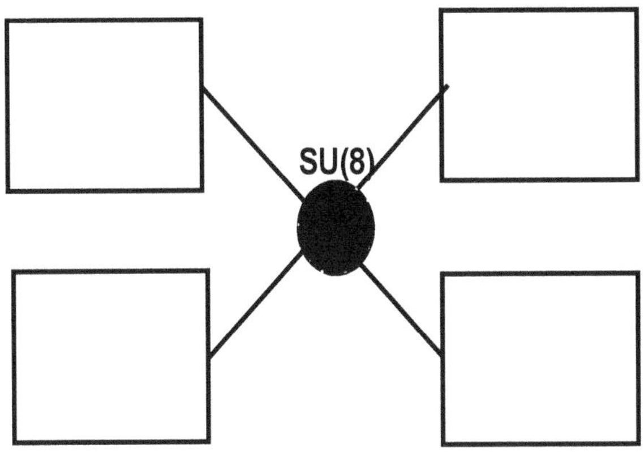

Figure 15-A.5. The SU(8) Connection Group of the Maxiverse connecting fermions in the four Megaverse "copies" blocks.

16. Comments on the Fermion Spectrum with Strata and Gambols

Earlier we defined the fundamental fermions and their mass spectrum by a map from a dimension array. Then we introduced four[44] strata of fermions that are exact replicates of the initial set of fundamental fermions differing only in masses. Stratum by stratum all masses increase by the multiple 95939. Subsequently we mapped each stratum of charged fundamental fermions to gambol fermions. The gambol fermion corresponding to each fermion of a stratum has 1/844242 of the fermions mass and 1/844242 of the fermion's symmetries additive quantum numbers. The currently known fundamental fermions were in stratum 1 or stratum 2.

We extended the sets of symmetry groups based on the nature of the strata and gambols. We now discuss some aspects of the combined fermion spectrum.

16.1 Form of Fermion Spectra

Fig. 16.1 shows the Normal sector fermions first stratum. This stratum has the neutrinos arranged in 4 generations. The fourth generation is as yet unknown.

Stratum 1

Layer 1:	Generation
ν_e e_1 u_1 d_1 p_1 s_1 c_1 b_1 t_1	1
ν_μ ...	2
ν_τ ...	3
$\nu_?$ ...	4
Plus 3 more layers	

Figure 16.1. The first (and lowest mass) UST stratum. It has 4 layers. The charged fermions have the subscript 1 indicating their stratum. The neutrinos, except $\nu_?$, are known neutrinos. The charged fermions have not yet been found. Their masses are a 1/95939 factor of the known charged fermions. The set of Dark sector fermions have the same form and are not shown.

The second UST stratum is depicted in Fig. 16.2. The neutrinos of the second strata are apparently unconfirmed. The charged fermions are known.

[44] The number of strata was chosen to be four for several reasons. First because of the "fourness" of generations and layers made it seem reasonable. Secondly, because the size of the multiplier 95939 causes a 5th strata, and higher number strata, fermions to have extremely large masses that would make them theoretically and experimentally separate Physically from the four strata that we selected.

Stratum 2

Layer 1:	Generation
ν_e' e u d s c b t	1
ν_μ' ...	2
ν_τ' ...	3
$\nu_?'$ ...	4

Plus 3 more layers

Figure 16.2. The second (and lowest mass) UST stratum. It also has 4 layers. Each layer has 4 generations. Each generation has 8 fermions plus a set of Dark fermions. The first generation of the first layer is depicted in detail. The fermions of the other generations and layers have an analogous form. Only Normal sector fermions are shown. The set of Dark sector fermions have the same form and are not shown. The second stratum fermions mirror those of the first stratum with the same quantum numbers but much larger masses due to the multiplier 95939.

16.2 The Gambol Fermions of Strata

Each charged fermion of each stratum has a corresponding gambol fermion with 1/844242 of the fermion's mass and 1/844242 of the additive quantum numbers of the fermion's additive quantum numbers.

Each charged fermion of each stratum can be viewed as equivalent[45] to 844242 gambols of corresponding type. The total masses and additive quantum numbers equal those of the fermion.

Each stratum has a set of gambols corresponding, on a one to one basis, with the charged fermions of the stratum. The gambols of one stratum can be viewed as composed of gambols of a lower stratum.

For example, strata 2, 3 and 4 gambols may contain corresponding gambols of strata below their stratum.

A charged stratum 4 fermion may contain some assemblage of gambols from strata 1, 2, and 3. A charged stratum 3 fermion may contain some assemblage of gambols from strata 1 and 2. The close correspondence of fermion and gambol properties from stratum to stratum makes gamboling possible. The confinement of gambols in fermions in each stratum beyond stratum 1 seems to be required – modulo possible *very* energetic interactions that liberate gambols.

We provisionally assume all neutrinos of all strata are *not* composed of gambols. The reason is the QED-like U(1) interaction that seems to be responsible for the existence of gambols as we saw in section 11.8. Gamboling is restricted to charged fermions.

The UST spectrum with strata and gambols that we have created can be directly extended to higher dimension spaces and universes.

---

[45] We discuss equivalence in the context of Lagrangians later.

The overall form of the strata in the UST displays a remarkable *fourfold* sequence of symmetry:

4-plets of up-type and down-type fermions (16.1)
4 generations
4 layers
4 strata

Fourfoldness can be extended to higher dimension spaces and universes.

16.3 Fermion Quadruplicates

Each fermion of a universe has four quadruplicates. The choice of quadruplicate number as 4 is additionally based on several points of Cosmos spaces' dimension array structure:

1. As shown in our earlier work dimension arrays are to symmetry groups and particles as Dirac matrices are to space-times. Both types of arrays are square. Both types of arrays relate directly to dimension through totally antisymmetric tensor properties. By choosing the quadruplication based on 4 we have the augmented dimension array for dimension r of 2^{r+6}. It is square and is the same as structurally as the dimension array for dimension r + 2. As a result we have a consistent formulation. (We note that the augmented r dimension array is only associated with the 2^{r+4} set of symmetries. Each stratum's fermions have the same set of symmetries. The r + 2 dimension array corresponds to a much larger set of symmetries.)

2. The choice of four replicates for each fermion is analogous to our use of quadplex fermion fields in Blaha (2024ℓ) to handle ElectroWeak and Strong interactions. Thus there is a consistency in formulation.

The Key Points of quadruplication is:

A. Each set of fermion quadruplicates has identical quantum numbers but differ in mass.

B. The fermions of one stratum can be treated as composed of corresponding gambol fermions of the same or lower strata.

17. Quad-Quad Equations for the First Generation of the First Layer[46]

We now consider UST interactions and dynamic equations using Quad-Quad[47] equations for the 1st generation of the 1st layer of the basic form of the UST without strata or gambols. The considerations in this chapter can be applied to other generations, layers, and strata (next chapter) by viewing them as composed of sets of 8 fermions and their corresponding symmetry groups.

In Blaha (2025a) we developed a new solution for White Holes and White Holes of the Second Kind. In addition we examined invariant intervals for subluminal and superluminal motion (for a constant force) and found that the intervals can be specified in the same form with an expression that uses the absolute value of the Lorentz factor

$$|\gamma| = (|1 - v^2/c^2|)^{-1/2} \qquad (17.1)$$

The subluminal and superluminal invariant interval expression was found to be:

$$\tau = t/|\gamma| = (mv_c/2F)\,[|\gamma|v/v_c + \text{arsinh}(|\gamma|v/v_c)/|\gamma|] \qquad (17.2)$$

where $|\gamma|$ is the absolute value of γ, F is a constant force, and v_c is the speed of light.

As a result we suggested a covariant superluminal and subluminal force law that extends Newton's Second Law to include superluminal motion:

$$F^\mu = m|\gamma|d(|\gamma|v^\mu)/dt \qquad (17.3)$$

We have extended this combined treatment of dynamics to quantum fields. We defined field operators that embodied both subluminal and superluminal dynamics in fields that we called SBT fields (SuperBaryonTachyon fields) $\psi_{SBT}(x)$. *The basic expressions in eq. 17.18 use* $|\gamma|$ *to implement a combined treatment of quantum field parts.*

The superluminal part of $\psi_{SBT}(x)$ has superluminal spinors and creation/annihilation operators. (To have canonical equal time anticommutation relations it is still necessary to use a light front coordinate system for the tachyonic part.) As we said in 2006:

> *The only approach that does succeed*[48] *is to decompose the tachyon field into left-handed and right-handed parts and then second quantize in light-front coordinates.*

---

[46] Blaha (2025c).

[47] Much of this chapter is derived from Blaha (2018e) and (2025c) as well as our other books.

[48] Blaha (2006) discusses this case in detail.

The light front representation of $\psi_{SBT}(x)$ quantum behavior is described in detail in chapter 3 of Blaha (2024j) and in (2006) and (2018e).

We find it necessary to separate momentum into velocity and Lorentz factor γ parts:

$$p^{\mu} = \gamma m v^{\mu} \tag{17.4}$$

The Fourier expansion of $\psi_{SBT}(x)$ extends to velocities from 0 to ∞:

$$\psi_{SBT}(x) = \sum_{s} \int_{0}^{\infty} dv d\Omega\, |p(v)|^2 J(|p|,v) N(|p|)[b_{SBT}(p, s)u_{SBT}(p, s)e^{-i|p|\cdot x} + d_{SBT}^{\dagger}(p, s)v_{SBT}(p, s)e^{+i|p|\cdot x}] \tag{17.5}$$

Note the integral extends from 0 to ∞ and thus encompasses both subluminal and superluminal velocities. Note also $\mathbf{p} = \mathbf{v}|p|$ and the Jacobian

$$J(|p|, v) = dp/dv = m|\gamma| + m(|\gamma|^6 - |\gamma|)^{1/2} v/c \tag{17.6}$$

17.1 Application to Sequences of Quadplex Fermion Fields

Blaha (2024j) introduced a new approach to The Unified SuperStandard Theory's (UST) Lorentz group $SO^+(1, 3)$ and its ElectroWeak SU(2)⊗U(1) and Strong Interaction SU(4) symmetries.[49] The starting point is the grouping of SU(2)⊗U(1) and $SO^+(1, 3)$ in the UST dimension array for our universe where $SO^+(1,3)$ has the four dimension coordinate system denoted y and SU(2)⊗U(1) has a real four dimension coordinate system denoted z. The z coordinate system supports the SU(2)⊗U(1) irreducible representation. The y coordinate system is the "real" coordinate system (our space-time) that we experience. We call it the LAB coordinate system.

Then SU(4) is introduced to create a unified set of interactions. SU(4) has two real four dimension coordinate systems denoted u and v supporting an SU(4) complex fundamental representation.

Two sequences of four fermions (whose masses were previously shown to form sequences) are introduced in section 17.2. They are the simultaneous basis of group representations of SU(2)⊗U(1) and SU(4). The sequences are used in the dynamic equations embodied in eq. 17.7. Representations of the fermions in terms of bradyon (b) and tachyon (t) parts of the fermion field operators are indicated in section 17.2.

A somewhat remarkable feature of the approach is the separation of the fermions by their bradyon and tachyon parts as indicated in section 17.2. *The separate representations into b and t parts is "just enough" to handle the sets of four fermions.*

Tachyon fields were embedded in duplex and quadplex fermion field formulations described in chapters 5 – 9 of Blaha (2924j). They are one application of our light front quantization described in the above book.

---

[49] An initial effort at this formulation appears in Blaha (2018e) where the ElectroWeak sector was shown to mirror the 4 dimension coordinate system and the Strong sector was also shown to be similar to the coordinate system suitably interpreted.

Earlier in section 9.2 of Blaha (2924j) we developed a UST Quadruplicated, Quadplexed Lagrangian for the Strong Interaction sector. We now have an approach that eliminates the differences between the bradyon and tachyon parts in the Lagrangian. As a result we can reexpress its eq. 9.12 without an $\mathcal{H}$ matrix as:

$$\begin{aligned}\mathcal{L} = &\overline{\psi}_{SBT\uparrow 2pa}{}^{\alpha\beta\rho\sigma}[M_{\uparrow a}{}^{-3}\gamma_{y\alpha\kappa}{}^{\mu}\partial/\partial y^{\mu}\,\gamma_{z\beta\lambda}{}^{\nu}\partial/\partial z^{\nu}\,\gamma_{u\rho\tau}{}^{\mu}\partial/\partial u^{\mu}\,\gamma_{v\sigma\chi}{}^{\nu}\partial/\partial v^{\nu} - M_{\uparrow a}]\psi_{SBT\uparrow 1pa}{}^{\kappa\lambda\tau\chi} + \\ &+ \overline{\psi}_{SBT\uparrow 1pa}{}^{\alpha\beta\rho\sigma}[M_{\uparrow a}{}^{-3}\gamma_{y\alpha\kappa}{}^{\mu}\partial/\partial y^{\mu}\,\gamma_{z\beta\lambda}{}^{\nu}\partial/\partial z^{\nu}\gamma_{u\rho\tau}{}^{\mu}\partial/\partial u^{\mu}\,\gamma_{v\sigma\chi}{}^{\nu}\partial/\partial v^{\nu} - M_{\uparrow a}]\psi_{SBT\uparrow 2pa}{}^{\kappa\lambda\tau\chi} + \\ &+ \overline{\psi}_{SBT\downarrow 2pa}{}^{\alpha\beta\rho\sigma}[M_{\downarrow a}{}^{-3}\gamma_{y\alpha\kappa}{}^{\mu}\partial/\partial y^{\mu}\,\gamma_{z\beta\lambda}{}^{\nu}\partial/\partial z^{\nu}\,\gamma_{u\rho\tau}{}^{\mu}\partial/\partial u^{\mu}\,\gamma_{v\sigma\chi}{}^{\nu}\partial/\partial v^{\nu} - M_{\downarrow a}]\psi_{SBT\downarrow 1pa}{}^{\kappa\lambda\tau\chi} + \\ &+ \overline{\psi}_{SBT\downarrow 1pa}{}^{\alpha\beta\rho\sigma}[M_{\downarrow a}{}^{-3}\gamma_{y\alpha\kappa}{}^{\mu}\partial/\partial y^{\mu}\,\gamma_{z\beta\lambda}{}^{\nu}\partial/\partial z^{\nu}\,\gamma_{u\rho\tau}{}^{\mu}\partial/\partial u^{\mu}\,\gamma_{v\sigma\chi}{}^{\nu}\partial/\partial v^{\nu} - M_{\downarrow a}]\psi_{SBT\downarrow 2pa}{}^{\kappa\lambda\tau\chi}\end{aligned} \tag{17.7}$$

where ψ_{SBTki} are PseudoQuantum field column vectors with k = up ↑ or down ↓ type fermions and i = 1, 2 specify the PseudoQuantum type of field, where p designates each of the fermions in a sequence, and where $\gamma_y{}^{\mu}$, $\gamma_z{}^{\mu}$, $\gamma_u{}^{\mu}$ and $\gamma_v{}^{\mu}$ are Dirac matrices for the y, z, u and v coordinates respectively. $M_{\uparrow}$ and $M_{\downarrow}$ are the 4×4 mass diagonal matrices of the up-type and down-type fermion sequences respectively. The sets of indices: αβρσ and κλτχ are spinor indices.

The *free* field equation is

$$[\gamma_{y\alpha\kappa}{}^{\mu}\partial/\partial y^{\mu}\,\gamma_{z\beta\lambda}{}^{\nu}\partial/\partial z^{\nu}\,\gamma_{u\rho\tau}{}^{\mu}\partial/\partial u^{\mu}\,\gamma_{v\sigma\chi}{}^{\nu}\partial/\partial v^{\nu} - M_{\uparrow a}{}^{4}]\psi_{SBT\uparrow kpa}{}^{\kappa\lambda\tau\chi} \tag{17.7a}$$

and similarly for the ↓ fermions where k = 1, 2 for the PseudoQuantum fields.

Eq. 17.7a can be separated into four equations. See eq. 17.29 below. The bradyon – tachyon distinction arises in the particle states.

17.2 SBT Fermion States

The states of the quadplex field, and their map to SU(4) fermions,[50] was shown to have the form

Up-Type Sequence

$$\begin{aligned} e &\leftrightarrow \psi_e = \psi_{1\uparrow} \sim b_y b_z t_u t_v \\ u &\leftrightarrow \psi_u = \psi_{2\uparrow} \sim b_y t_z t_u b_v \\ c &\leftrightarrow \psi_c = \psi_{3\uparrow} \sim b_y b_z b_u t_v \\ t &\leftrightarrow \psi_t = \psi_{4\uparrow} \sim b_y b_z b_u b_v \end{aligned} \tag{17.8}$$

Down-Type Sequence

$$\begin{aligned} \nu_e &\leftrightarrow \psi_e = \psi_{1\downarrow} \sim t_y t_z t_u t_v \\ \nu' &\leftrightarrow \psi_{\nu'} = \psi_{1\downarrow} \sim b_y t_z t_u t_v \\ d &\leftrightarrow \psi_d = \psi_{2\downarrow} \sim b_y t_z t_u b_v \\ s &\leftrightarrow \psi_s = \psi_{3\downarrow} \sim b_y t_z b_u t_v \\ b &\leftrightarrow \psi_b = \psi_{4\downarrow} \sim b_y t_z b_u b_v \end{aligned}$$

[50] This discussion applies to SU(3)⊗U(1) with little change.

where we use particle symbols to represent wave functions with the coordinate systems indicated with subscripts. Fields have four parts of either the bradyon, denoted b, or tachyon, denoted t, type.

Based on the above we can identify SBT states directly. For example, the state corresponding to the u quark is

$$| \,p1,s_1;\, q2,s_2;\, q3,s_3;\, p4,s_4> \qquad (17.9)$$

and its complex conjugate is

$$< p1,s_1;\, q2,s_2;\, q3,s_3;\, p4,s_4| \qquad (17.10)$$

where p1 and p4 are subluminal momenta and q2 and q3 are superluminal momenta.

With this type of mapping we use the above free Lagrangian in eq. 17.7 augmented by interactions, and calculate amplitudes using a quadplex LSZ expansion in the four coordinate systems and their conjugate momentums.

Thereby we have moved the dependence of the Lagrangian on the difference between bradyon and tachyon fields to its set of states.

17.3 Generating States with Operators for a Quantum Field

We begin with the features of a quantum field embodying subluminal and superluminal parts. In section 17.4 we describe the quadplex quantum field with the form of eq. 17.5.

17.3.1 Subluminal Part

The form of the spinor solutions u(p, s) and v(p, s) in the subluminal case can be determined from the spinor Lorentz group transformation. Defining

$$\cosh(\omega/2) = [(E+m)/(2m)]^{½} \text{ and } \sinh(\omega/2) = |\mathbf{p}|[2m(E+m)]^{-½} \qquad (17.11)$$

we find the subluminal form:

$$S_L(\Lambda(\omega, \mathbf{u})) = \begin{array}{cccc} \mathbf{u(p, s,v)} & \mathbf{u(p, -s,v)} & \mathbf{v(p, s,v)} & \mathbf{v(p, -s)} \\ \left[\begin{array}{cccc} \cosh(\omega/2) & 0 & \sinh(\omega/2)p_z/p & \sinh(\omega/2)p_-/p \\ 0 & \cosh(\omega/2) & \sinh(\omega/2)p_+/p & -\sinh(\omega/2)p_z/p \\ \sinh(\omega/2)p_z/p & \sinh(\omega/2)p_-/p & \cosh(\omega/2) & 0 \\ \sinh(\omega/2)p_+/p & -\sinh(\omega/2)p_z/p & 0 & \cosh(\omega/2) \end{array}\right] \end{array} \quad (17.12)$$

where v is the magnitude of the velocity.

17.3.2 SUPERLUMINAL PART

The form of the spinor solutions u(p, s, v) and v(p, s, v) in the superluminal case are also determined from the spinor Lorentz group transformation. Let

$$\cosh(\omega_L/2) = [(iE+m)/(2m)]^{½} \text{ and } \sinh(\omega_L/2) = |\mathbf{ip}|[2m(iE+m)]^{-½} \quad (17.13)$$

or

$$\cosh(\omega_L/2) = [(i\gamma mv^0+m)/(2m)]^{½} \text{ and } \sinh(\omega_L/2) = i\gamma|\mathbf{v}|[2m(i\gamma mv^0+m)]^{-½} \quad (17.14)$$

since $p^\mu \rightarrow ip^\mu$ transitioning from subluminal to superluminal and using eq. 17.4. Defining $\gamma_s = i\gamma = (\beta^2 - 1)^{-½}$ we obtain

$$\cosh(\omega_L/2) = [(\gamma_s v^0+1)/2]^{½} \quad (17.15a)$$

and

$$\sinh(\omega_L/2) = \gamma_s|\mathbf{v}|[2(\gamma_s v^0+1)]^{-½} \quad (17.15b)$$

The superluminal spinor solutions are the columns:

$$S_L(\Lambda_L(\omega, \mathbf{u})) = \begin{array}{cccc} \mathbf{v_T(p, s,v)} & \mathbf{v_T(p, -s,v)} & \mathbf{u_T(p, s,v)} & \mathbf{u_T(p, -s,v)} \\ \left[\begin{array}{cccc} \cosh(\omega_L/2) & 0 & \sinh(\omega_L/2)p_z/p & \sinh(\omega_L/2)p_-/p \\ 0 & \cosh(\omega_L/2) & \sinh(\omega_L/2)p_+/p & -\sinh(\omega_L/2)p_z/p \\ \sinh(\omega_L/2)p_z/p & \sinh(\omega_L/2)p_-/p & \cosh(\omega_L/2) & 0 \\ \sinh(\omega_L/2)p_+/p & -\sinh(\omega_L/2)p_z/p & 0 & \cosh(\omega_L/2) \end{array}\right] \end{array} \quad (17.16)$$

where v is the magnitude of the velocity. See Appendix 2-A of Blaha (2018e) for more details.

17.3.3 SBT SPINOR SOLUTIONS FOR THE SUBLUMINAL AND SUPERLUMINAL

We define $u_{SBT}(p, s)$ and $v_{SBT}(p, s)$ as

$$u_{SBT}(p,s,v) = u(p, s,v)\theta(c - v) + v_T(p, s,v)\theta(v - c) \tag{17.17a}$$
$$u_{SBT}(p, -s,v) = u(p, -s,v)\theta(c - v) + v_T(p, -s,v)\theta(v - c) \tag{17.17b}$$

where c is the speed of light and where v is the magnitude of the velocity. using

$$\cosh(\omega_L/2) = [(|\gamma|v^0+1)/2]^{1/2} \tag{17.18a}$$
$$\sinh(\omega_L/2) = |\gamma||\mathbf{v}|[2(|\gamma|v^0+1)]^{-1/2} \tag{17.18b}$$

and

$$v_{SBT}(p, s,v) = v(p, s,v)\theta(c - v) + u_T(p, s,v)\theta(v - c) \tag{17.19a}$$
$$v_{SBT}(p, -s,v) = v(p, -s,v)\theta(c - v) + u_T(p, -s,v)\theta(v - c) \tag{17.19b}$$

17.3.4 RAISING AND LOWERING OPERATORS FOR THE SUBLUMINAL AND SUPERLUMINAL

We can define raising and lowering operators for the combined subluminal and superluminal operators.

Subluminal Operator Anticommutaion Relations

$$\{b(q,s,v), b^\dagger(p,s',v)\} = \delta_{ss'}\delta^3(\mathbf{q} - \mathbf{p}) \tag{17.20}$$
$$\{d(q,s,v), d^\dagger(p,s',v)\} = \delta_{ss'}\delta^3(\mathbf{q} - \mathbf{p})$$

with the other combinations zero.

Superluminal Operator Anticommutaion Relations

These operator anticimmutators must be expressed in terms of light front coordinates. For example, the left-handed, "+" light front non-zero anticommutators are:[51]

$$\{b_{TL}{}^+(q,s,v), b_{TL}{}^{+\dagger}(p,s',v)\} = \delta_{ss'}\delta^2(\mathbf{q} - \mathbf{p})\delta(q^+ - p^+) \tag{17.21}$$
$$\{d_{TL}{}^+(q,s,v), d_{TL}{}^{+\dagger}(p,s',v)\} = \delta_{ss'}\delta^2(\mathbf{q} - \mathbf{p})\delta(q^+ - p^+)$$

Combined Operators

$$b_{SBT}(p, s,v) = b(p, s,v)\theta(c - v) + b_{TL}{}^+(p, s,v)\theta(v - c) \tag{17.22}$$
$$b_{SBT}{}^\dagger(p, s,v) = b^\dagger(p, s,v)\theta(c - v) + b_{TL}{}^{+\dagger}(p, s,v)\theta(v - c)$$
$$d_{SBT}(p, s,v) = d(p, s,v)\theta(c - v) + d_{TL}{}^+(p, s,v)\theta(v - c)$$
$$d_{SBT}{}^\dagger(p, s,v) = d^\dagger(p, s,v)\theta(c - v) + d_{TL}{}^{+\dagger}(p, s,v)\theta(v - c)$$

The anticommutators of the combined operators in eq. 17.22 give the results of eqs. 17.20 and 17.21 for $v < c$ and $v > c$ respectively. The anticommutaors are

$$\{ b_{SBT}(p, s, v), b_{SBT}{}^\dagger(p', s', v)\} = \delta_{ss'}\delta^3(\mathbf{p} - \mathbf{p'}, v) \tag{17.22a}$$

[51] From Blaha (2018e) section 2.4.

$$\{ d_{SBT}(p, s, v), d_{SBT}^{\dagger}(p', s', v)\} = \delta_{ss'}\delta^3(\mathbf{p} - \mathbf{p'}, v)$$

with other anticommutators being zeros, where we define a combined delta function for subluminal and for superluminal (left-handed + light front):

$$\delta^3(\mathbf{p}, v) = \theta(c - v)\delta^3(\mathbf{p}) + \theta(v - c)\,\delta^2(\mathbf{p})\delta(p^+) \qquad (17.22b)$$

17.3.5 EXTRACTING THE OPERATORS FOR THE SUBLUMINAL AND SUPERLUMINAL

We can project operators from the field operator using:

Subluminal

$$U_{sub\alpha}(p, s, x,v) = \theta(c - v)(2\pi)^{-3/2}\,(m/E)^{1/2}\,e^{-ip.x}\,u_{SBT\alpha}(p, s,v) \qquad (17.23)$$

$$\theta(c - v)\,b^{\dagger}(p, s,v) = \int d^3x\,\psi_{SBT\alpha}^{\dagger}(x)U_{sub\alpha}(p, s, x,v) \qquad (17.24)$$

Superluminal

$$U_{super\alpha}(p, s, x,v) = \theta(v - c)\,[2m|\mathbf{p}|/((2\pi)^3(p^+(p^+ - p^-) + p_\perp^{\,2}))]^{1/2}\,e^{-ip.x}\,u_{SBT\alpha}(p, s,v) \qquad (17.25)$$

$$\theta(v - c)b_{TL}^{\;+\dagger}(p, s,v) = \int d^3x\,\psi_{SBT\alpha}^{\dagger}(x)U_{super\alpha}(p, s, x,v) \qquad (17.26)$$

using $[2m|\mathbf{p}|/((2\pi)^3(p^+(p^+ - p^-) + p_\perp^{\,2}))]^{1/2}$ from eq. 2.73 in Blaha (2018e).

Other operator extractions have corresponding expressions. In general

$$U_\alpha(p, s, x, v) = \theta(c - v)\,U_{sub\alpha}(p, s, x,v) + \theta(v - c)U_{super\alpha}(p, s, x,v) \qquad (17.27)$$

17.3.6 EXTRACTING PARTICLE STATES

We can now use eqs. 17.24 and 17.26 to form particle states. For example, a bradyon state ($v < c$)

$$|p, s>_{brady} = b^{\dagger}(p, s,v)|0> \qquad (17.28)$$

and a left-handed "+" light front tachyon state ($v > c$) is

$$|p, s>_{tachy} = b_{TL}^{\;+\dagger}(p, s,v)|0> \qquad (17.29)$$

17.4 Quadplex Field

In this section we describe the quadplex quantum field with the form of eq. 17.5. The generation of quadplex states is based on interaction of the above operator expressions.

Upon variation and separation of equations we obtain:

$$\begin{aligned}
&[i\gamma_y^{\mu}\partial/\partial y^{\mu} - M]\psi = 0 \\
&[i\gamma_z^{\mu}\partial/\partial z^{\mu} - M]\psi = 0 \\
&[i\gamma_u^{\mu}\partial/\partial u^{\mu} - M]\psi = 0 \\
&[i\gamma_v^{\mu}\partial/\partial v^{\mu} - M]\psi = 0
\end{aligned} \qquad (17.30)$$

Thus $\psi = \psi_{SBTki}$ for k = 1(up type) or 2 (down type), and i = 1, 2, is

$$\psi_{SBTki\alpha\beta\gamma\delta}(y,z,u,v) = \sum_{s_1}\sum_{s_2}\sum_{s_3}\sum_{s_4} \int dq_1^{r-1}dq_2^{r-1}dq_3^{r-1}dq_4^{r-1}N(q_1,q_2,q_3,q_4)[b_i(q_1, s_1,v_y,q_2,s_2,v_z,q_3,s_3, v_u,q_4,s_4,v_v) \cdot u_\alpha(q_1,s_1,v_y)\, u_\beta(q_2,s_2,v_z)u_\gamma(q_3,s_3,v_u)u_\delta(q_4,s_4,v_v) \cdot \cdot\exp(-iq_1\cdot y - iq_2\cdot z - iq_3\cdot u - iq_4\cdot v) + + d^\dagger_i(q_1,s_1,v_y,q_2,s_2,v_z,q_3,s_3,v_u,q_4,s_4,v_v) \cdot \cdot\, v_\alpha(q_1,s_1,v_y)v_\beta(q_2, s_2,v_z)v_\gamma(q_3,s_3,v_u)v_\delta(q_4, s_4,v_v) \cdot \cdot\, \exp(iq_1\cdot y + iq_2\cdot z + iq_3\cdot u + iq_4\cdot v)] \quad (17.31)$$

where r is the space-time dimension 4, and where $N(q_1,q_2,q_3,q_4)$ is a normalization factor. Note: all parts (in y, z, u, and v) of ψ_{SBTki} are on the mass shell: $q_i^2 = M^2$ for i = 1,2,3,4. Note the spinor functions are denoted u and v. The v functions are not related to the variable v.

17.4.1 Extracting Creation/Annihilation Operators

From eq. 17.27 we obtain the projections into operators:

$$b_k^\dagger(q_1,s_1,v_y,q_2,s_2,v_z,q_3,s_3,v_u,q_4,s_4,v_v) = N(q_1,q_2,q_3,q_4)\int dy^{r-1}dz^{r-1}du^{r-1}dv^{r-1}\psi_{SBTk2\alpha\beta\gamma\delta}{}^\dagger(y,z,u,v) \cdot \cdot U_\alpha(q_1,s_1,y,v_y)U_\beta(q_2,s_2,z,v_z)U_\gamma(q_3,s_3,u,v_u)U_\delta(q_4,s_4,v,v_v) \quad (17.32)$$

$$d_k^\dagger(q_1,s_1,v_y,q_2,s_2,v_z,q_3,s_3,v_u,q_4,s_4,v_v) = N(q_1,q_2,q_3,q_4)\int dy^{r-1}dz^{r-1}du^{r-1}dv^{r-1} \cdot \cdot V_\alpha^\dagger(q_1,s_1,y,v_y)V_\beta^\dagger(q_2,s_2,z,v_z)V_\gamma^\dagger(q_3,s_3,u,v_u)V_\delta^\dagger(q_4,s_4,v,v_v)\psi_{SBTk2\alpha\beta\gamma\delta}(y,z,u,v) \quad (17.33)$$

using a type 2 PseudoQuantum field for extracting a field operator as we have done consistently in the past. The subscript k represents ↑ up-type and ↓ down-type fermion (See Fig. 17.8.). Other operator extractions have similar expressions.
The non-zero anticommutators are:

$$\{b_k(q_1,s_1,v_y,q_2,s_2,v_z,q_3,s_3,v_u,q_4,s_4,v_v),\, b_k^\dagger(q_1',s_1',v_y,q_2',s_2',v_z,q_3',s_3',v_u,q_4',s_4',v_v)\} = = \delta_{s_1s_1'}\delta^3(\mathbf{q}_1 - \mathbf{q}_1', v_y)\, \delta_{s_2s_2'}\delta^3(\mathbf{q}_2 - \mathbf{q}_2', v_z)\, \delta_{s_3s_3'}\delta^3(\mathbf{q}_3 - \mathbf{q}_3', v_u)\, \delta_{s_4s_4'}\delta^3(\mathbf{q}_4 - \mathbf{q}_4', v_v) \quad (17.34)$$

$$\{d_k(q_1,s_1,v_y,q_2,s_2,v_z,q_3,s_3,v_u,q_4,s_4,v_v),\, d_k^\dagger(q_1',s_1',v_y,q_2',s_2',v_z,q_3',s_3',v_u,q_4',s_4',v_v)\} = = \delta_{s_1s_1'}\delta^3(\mathbf{q}_1 - \mathbf{q}_1', v_y)\, \delta_{s_2s_2'}\delta^3(\mathbf{q}_2 - \mathbf{q}_2', v_z)\, \delta_{s_3s_3'}\delta^3(\mathbf{q}_3 - \mathbf{q}_3', v_u)\, \delta_{s_4s_4'}\delta^3(\mathbf{q}_4 - \mathbf{q}_4', v_v) \quad (17.35)$$

where k specifies up-type or down-type and where we use the combined delta function for subluminal and for superluminal (left-handed + light front) in eq. 17.22b.

17.4.2 Defining States

We can define single quadplex particle states using eq. 17.32. For example the proton state represented in Fig. 7.8:

$$u \leftrightarrow \psi_u = \psi_{2\uparrow} \sim b_y t_z t_u b_v$$

has bradyon y (LAB coordinates), tachyon z, tachyon u, and bradyon v parts:

$$|q_1,s_1,v_y;q_2,s_2,v_z;q_3,s_3,v_u;q_4,s_4,v_v> = b_k^\dagger(q_1,s_1,v_y,q_2,s_2,v_z,q_3,s_3,v_u,q_4,s_4,v_v)|0> \quad (17.36)$$

where $v_y < c$, $v_z > c$, $v_u > c$, and $v_v < c$.

Thus we can represent fermions in this quadplex formulation with the bradyon vs. tachyon nature of the parts deferred until the definition of states.

We can create a particle superposition with

$$|\text{particle}> = N\int dq_1^{r-1}dq_2^{r-1}dq_3^{r-1}dq_4^{r-1}\ f(q_1,\ q_2,q_3,q_4)\ |q_1,s_1,v_y;q_2,s_2,v_z;q_3,s_3,v_u;q_4,s_4,v_v> \quad (17.37)$$

Multiparticle states can be defined directly.

These types of states can be used in an LSZ formulation of perturbation theory.

18. Quadruplicated, Quadplexed Lagrangian Extended to all Generations and Layers

18.1 Extension of UST Quadruplicated, Quadplexed Lagrangian to All Generations and Layers of One Stratum

The Quadruplicated, Quadplexed Lagrangian described in the previous chapter applies to the first generation of the first layer of fundamental fermions. We can extend the form of that Lagrangian to the Dark Matter first generation and layer directly modulo the values of masses and coupling constants.

The previous chapter gives the Lagrangian for one generation of one layer of one stratum. It directly generalizes to the four generations of a layer. To achieve this goal the set of fermions must first be expanded to the fermions of all generations of a layer. Their interactions (described below) generalize directly.

Then the Lagrangian must generalize to all layers of a stratum. *Here we must recognize the ElectroWeak groups, Strong Interaction groups, Generation groups, and Layer groups are different in each layer; and differ between Normal and Dark Matter.*

The interaction terms for the ElectroWeak, Strong, Generation[52] Group and Layer Group interactions are obtained by imposing gauge covariance. Then one finds the interactions between the fermions of a stratum follow from:

$$\begin{aligned}
\partial/\partial y^\mu &\rightarrow \partial/\partial y^\mu - ig_S A_S^{k\mu}(y)T_{Syk} - ig_W A_W^{k\mu}(y)T_{Wyk} - ig_G A_G^{k\mu}(y)T_{Gyk} - ig_L A_L^{k\mu}(y)T_{Lyk} - ig_{WO} A_{WO}^{k\mu}(y)T_{WOyk} \\
\partial/\partial z^\mu &\rightarrow \partial/\partial z^\mu - ig_S A_S^{k\mu}(z)T_{Szk} - ig_W A_W^{k\mu}(z)T_{Wzk} - ig_G A_G^{k\mu}(z)T_{Gzk} - ig_L A_L^{k\mu}(z)T_{Lzk} - ig_{WO} A_{WO}^{k\mu}(z)T_{WOzk} \\
\partial/\partial u^\mu &\rightarrow \partial/\partial u^\mu - ig_S A_S^{k\mu}(u)T_{Suk} - ig_W A_W^{k\mu}(u)T_{Wuk} - ig_G A_G^{k\mu}(u)T_{Guk} - ig_L A_L^{k\mu}(u)T_{Luk} - ig_{WO} A_{WO}^{k\mu}(u)T_{WOuk} \\
\partial/\partial v^\mu &\rightarrow \partial/\partial v^\mu - ig_S A_S^{k\mu}(v)T_{Svk} - ig_W A_W^{k\mu}(v)T_{Wvk} - ig_G A_G^{k\mu}(v)T_{Gvk} - ig_L A_L^{k\mu}(v)T_{Lvk} - ig_{WO} A_{WO}^{k\mu}(v)T_{WOvk}
\end{aligned} \tag{18.1}$$

where T_{IJk} is the kth generator matrix element for I = SU(4), SU(2), Generation U(4), Layer U(4), and ElectroWeak U(1) for the J = y. z, u, and v coordinate parts of the quadplex fermion field. Also g_I is the coupling constant.

Fig. 18.1 displays the interaction groups for one stratum. Note each group is for a specific layer with the exception of the four Layer Groups that cause interactions between fermions of differing layers. We grouped Layer groups within layers for convenience. There are 8 ElectroWeak groups, 8 Strong Interaction groups, 8 Generation groups, and 8 Layer groups. Their irreducible representations' dimensions add to 256 as indicated in Fig. 18.1.

The set of interaction groups must be extended to include the interactions of the Generation Groups and the Layer Groups since the Quadruplicated, Quadplexed

[52] The Generation Group and Layer group interactions must be introduced upon extension to all generations and layers of a stratum.

Lagrangian must be extended to all generations and layers of a stratum for completeness.

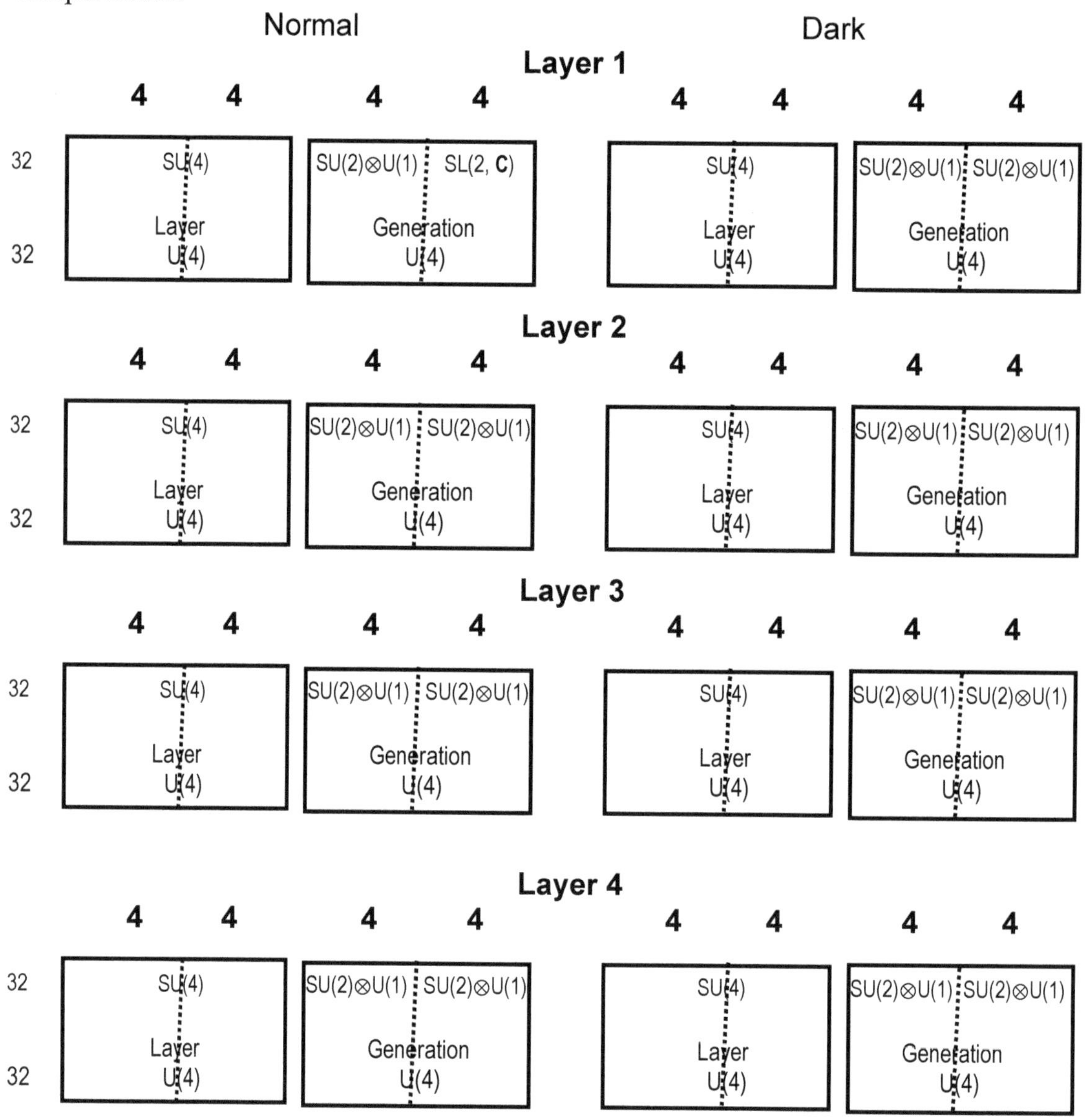

Figure 18.1. (From Blaha (2024i)). Normal and Dark symmetry groups of UST. SL(2, C) represents the Lorentz group SO^{+}(1,3). This diagram appears in Blaha (2024i) and our earlier books such as Blaha (2020d).

18.2 Extension of UST Quadruplicated, Quadplexed Lagrangian to All Strata

The previous section specifies the Lagrangian and interactions for one stratum. To incorporate all four strata we must do the following. The fermions of the one stratum must be augmented with the fermions of the other strata. In doing this we must assume that the symmetry groups and their interactions are the same for all strata. *The strata **share** the same set of groups and interactions.* Corresponding fermions of each stratum have the same quantum numbers. These requirements lead to the formulation for all four strata.

The generalization of the Lagrangian theory to strata as above results in:

1. All 1024 fundamental UST fermions of all strata are included.
2. The group structure and interactions are well defined.
3. The groups and interactions are the same: stratum by stratum.
4. The fermion interactions are strata independent.

In addition:

5. A Gambol Theory of fermions is supported. Four gambol fermion strata are supported as discussed earlier.

It is possible to create a variation on this theory that supports different interactions between the fermions in different strata. New stratum specific interactions could be introduced. One possibility is to define *Strata Connection Groups* that connect corresponding fermions in different strata as discussed in earlier chapters. Fig. 18.2 illustrates this possibility with six Strata Connection Groups. We provisionally assume these groups have ultra weak interactions similar in form to the SU(2)⊗U(1) ElectroWeak-like interactions.

18.3 Connection Groups Between Strata[53]

The four strata may be interconnected by a set of Connection groups. These group interactions may be visualized as in Fig. 18.2. An interaction between a fermion of one stratum is with a corresponding fermion of the other stratum. There are six possible Connection groups for interactions between fermions of four different strata. Each of the groups may be viewed as a very weak SU(2)⊗U(1) ElectroWeak-like interaction. Each of the 256 distinct fermions populating each stratum is a irreducible two dimension representation of an SU(2)⊗U(1) group. One expects the strength of these interactions is ultra weak since no Physical evidence has appeared.

18.4 Introduction of Gambols

We can introduce a set of gambol fermions for each fermion in each stratum. Each fermion has *one* corresponding gambol fermion. Any fermion can be "mapped" to a set of 844242 of its corresponding gambols. (Any or all of these gambols can also be expressed as corresponding gambols of a lower stratum.)

The mass of each gambol is a factor of 1/844242 of the mass of its corresponding fermion. The additive quantum numbers of each gambol fermion are also a factor of 1/844242 of the additive quantum numbers of the corresponding fermion.

The gambols of a fermion of one stratum can be viewed as composed of corresponding gambol fermions of the next lower stratum (excepting the first stratum). Gambol Theory as described in Blaha (2025c) is then applicable. We discussed Gambol Theory in detail in earlier books.

---

[53] Part of this section appears in Appendix 7-B from Blaha (2024ℓ) as part of a discussion of Connection Groups of the UST.

18.5 Generalization to the Universes of Other Spaces

The generalization of these strata features to universes of other higher dimension spaces is direct. One simply views the higher dimension universe as a quadruple (or a series of quadruples) of the structure of our universes' UST structure.

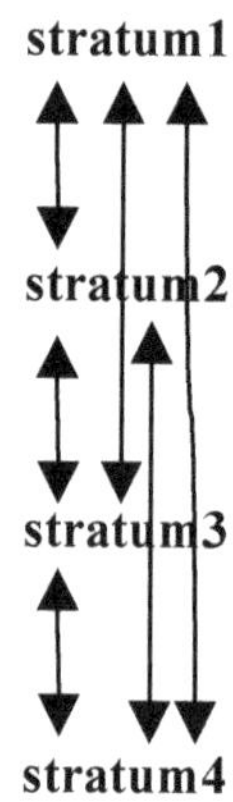

Figure 18.2. Strata Connection groups. A view of the four UST strata of the Sequence Theory showing the 6 groups and interactions relating the fermions in the various strata. The interaction indicated with the arrows is between a fermion of one stratum with the corresponding fermion of another stratum.

19. Gambol Constitiuents of Fermions

We have suggested the stratums of fermions can be combined with Gambol Theory. The fermions of each stratum may have gambol fermion constituents, These constituents can have "sun-constituents" which are fermion gambols of strata below them.

Each of the UST charged fundamental fermions has a corresponding gambol fermion. The neutral fermions, the neutrinos, do not have a corresponding gambol fermion as we showed earlier.

A *charged* fermion of a given stratum is equivalent[54] to 844242 constituent gambol fermions *corresponding* to its features and of the same stratum. These gambols have the same symmetry group identity as the fermion. However their masses and symmetry group additive quantum numbers are reduced by a factor of 844242.

When a charged fermion interacts the corresponding constituent gambols' additive quantum numbers sum to equal the fermion's additive quantum numbers. Similarly the 844242 gambol fermion constituent masses sum to the mass of the fermion.

The introduction of a gambol fermion for each charged fermion in each stratum significantly increases the total number of distinct UST fermions to

$$224 \times 2 \text{ states (fermions and gambols)} \times 4 \text{ strata} + 32 \times 4 \text{ strata} = 1920 \text{ fermions}$$

by eq. 11.16. Note neutrinos do not have corresponding gambols.

For a dimension r HyperCosmos space's universe the total number of dimension r fermions fermions (including gambols) is based on:

2^{r+4} fermions of which $2^{r+4} - 2^{r+1}$ are charged
2^{r+1} Normal and Dark neutrinos
$2^{r+4} - 2^{r+1}$ possible fermion and gambol pairs

implying The total number of dimension r fermions including gambols is:

$$(2^{r+4} - 2^{r+1}) \times 2 \text{ states} \times 4 \text{ strata} + 2^{r+1} \times 4 \text{ strata} = 2^{r+7} - 2^{r+4} + 2^{r+3} \text{ fermions}$$

from section 11.6.4.

The somewhat large number of gambols, 844242, equivalent to a fermion is less challenging when one realizes as we did in earlier books that the gambol constituents of an individual fermion or a set of individual fermions can be handled *statistically* as we

[54] Equivalence can take a number of forms. We chose to implement it with interaction terms. These interactions are described later.

did in earlier books such as Blaha (2023e) for particle decays and scattering. The very large numbers of gambols within a fermion makes statistical analysis practical.

The number of gambol constituents of a fundamental fermion ranges from

> 1 to 844242 if the constituents are all of the same stratum as the fermion One Can group these gambols together into sets of composite gambols
>
> Any number of constituent gambol fermions of the fermion's stratum can be "expanded" into a combination of gambols of the same stratum and of gambols of lower stratums.

Thus all is possible – The number of constituents of a fundamental fermion ranges from 1 to an extremely large number: For all gambols of the same stratum as the fermion – 844242. For all gambols of the same stratum as the fermion transformed to gambols of the next lower stratum – up to 844242^2. And so on. Thus our *statistical* Gambol Theory formulation is supported.

19.1 Gambol Constituents

The fundamental fermions may be viewed as having a dual identity. On the one hand we have each of the 1024 UST fermions' quantum numbers as the same for all four strata – fermion by fermion. They do differ by a stratum quantum number.

On the other hand, each fermion of the 896 UST charged fermions in strata has a corresponding set of 844242 gambol fermions "within" it with each gambol fermion's mass reduced by a factor of 844242 and with each gambol fermion's additive quantum numbers also reduced by a factor of 844242 from that of the corresponding fermion.

How is this dual identity implemented? We have suggested previously[55] that Lagrangian field theory models can be developed that implement this view of the fermion strata. We can implement this model with transition terms which specify transitions between each fermion and its set of 844242 gambols.[56] This set of transitions can be iterated to relate a fermion to sets of gambols within gambols.

By specifying each charged fundamental fermion (quarks and leptons and their combinations within hadrons) as having a set of gambols, the issue of gambol confinement surfaces. We will assume gambols are confined within fundamental fermions, and hadrons, and hadron interaction regions. The confinement mechanism then surfaces. One might suggest it is similar to quark confinement.

We suggest a new mechanism based on a restriction on the allowed asymptotic S-matrix states. We suggest gambols only occur within particles and interaction regions but not as particles in asymptotic states. This restriction on the set of asymptotic states

---

[55] See Chapter 5 of Blaha (2023e).

[56] Interactions with products of many fermion fields are renormalizable using our Two Tier formalism developed in the early 2000's.

of a theory which eliminates unwanted asymptotic states is not a new idea. For example such restrictions appear in certain formulations of QED.

19.2 Gambol Fermion Fields

We now turn to develop a relatively simple Lagrangian field theory for a single fermion with a set of associated gambols. We define PseudoQuantum free fermion field:

$$\psi_1(x) = \Sigma_\alpha\, [b_{1\alpha}\, f_\alpha(x) + d_{1\alpha}^{\dagger} f_\alpha^*(x)] \qquad (19.1)$$
$$\psi_2(x) = \Sigma_\alpha\, [b_{2\alpha} f_\alpha(x) + d_{2\alpha}^{\dagger}\, f_\alpha^*(x)]$$

where the $f_\alpha(x)$ are Fourier fields with α being the momentum, and the gambol PseudoQuantum free fermion fields:

$$\psi^g{}_1(x) = \Sigma_\alpha\, [b^g{}_{1\alpha}\, f_\alpha(x) + d^g{}_{1\alpha}{}^{\dagger} f_\alpha^*(x)] \qquad (19.2)$$
$$\psi^g{}_2(x) = \Sigma_\alpha\, [b^g{}_{2\alpha} f_\alpha(x) + d^g{}_{2\alpha}{}^{\dagger}\, f_\alpha^*(x)]$$

where the $f_\alpha(x)$ are Fourier fields with α being the momentum

In Blaha (2023e) we considered gambol fermion fields. We considered a fermion and formulated a set of n gambols equivalent to it. We developed the anti-commutation relations gambol fractionations 1/n and 1/m:

$$\{b^{1/n}(q),\, b^{1/m\dagger}(p)\} = \delta_{nm}\delta^{r-1}(q-p) \qquad (19.3)$$
$$\{b^{1/n}(q),\, b^{1/m}(p)\} = 0$$
$$\{b^{1/n\dagger}(q),\, b^{1/m\dagger}(p)\} = 0$$

and similarly for $d^{1/n}$ and $d^{1/n\dagger}$:

$$\{d^{1/n}(q),\, d^{1/m\dagger}(p)\} = \delta_{nm}\delta^{r-1}(q-p) \qquad (19.4)$$
$$\{d^{1/n}(q),\, d^{1/m}(p)\} = 0$$
$$\{d^{1/n\dagger}(q),\, d^{1/m\dagger}(p)\} = 0$$

Note: fractional gambol operators of different fractionation anti-commute.

We might introduce a Lagrangian interaction between gambols after setting g = 1/n in the above creation and annihilation operators:

$$\sum_{i=1}^{4} \sum_{j=1}^{4} \overline{\psi}_2{}^g{}_i(y) \sum_k A_k{}^\mu(y)\, \gamma_\mu T_{kij}\, \psi_2{}^g{}_j(y) \;+ \text{c.c.} \qquad (19.5)$$

where $A_k{}^{a\mu}$ is a gauge field, T_k is a group generator, and where we use type 2 PseudoQuantum fields as we have done since our quark confinement and quantum field theory papers in the 1970s.[57]

---

[57] These referreed papers by the author appeared in Phys. Rev. D and Il Nuovo Cimento in the 1970s. Among these papers are: S. Blaha, Phys. Rev. **D17**, 994 (1978).; _______, Phys. Rev. D**11**, 2921 (1975); ______, Phys. Rev. **D10**, 4268 (1974); ______, IL Nuovo Cimento **49A**, 35 (1979); and _____, IL Nuovo Cimento **49A**, 113 (1979).

The fermion-gambol interaction could have a mediating gauge field. We will assume there is no mediating gauge field connecting fermions and their gambols. There is apparently no evidence for it is in deep inelastic scattering and hadron decay cases.

We will therefore use the connecting interaction:

$$g\ \overline{\psi}_2(x)\prod_{n=1}^{844242}\psi_2{}^g(x) \tag{19.6}$$

Eq. 19.6 is a product of many fermion fields. In conventional perturbation theory the results would be non-renormalizable. However our Two Tier Formalism developed in the early 2000's gives full renormalizability. See Blaha (2002), *A Finite Unified Quantum Field Theory of the Elementary Particle Standard Model and Quantum Gravity Based on New Quantum Dimensions™ & a New Paradigm in the Calculus of Variations*. Thus this interaction can appear in a renormalizable theory using the author's Two Tier Formalism.[58]

The 844242-fold $\psi_2{}^g$ gambol factors in eq. 19.6 lead to Feynman graph terms of the form of Fig. 19.1 for the fermion propagator. We can interpret the impact of these terms as giving a free field fermion propagator at low energies and a set of gambol propagators to leading order at high energies. We may require the impact of these propagators to correspond quite like the data from deep inelastic lepton-nucleon scattering experiments.[59] The large number of gambol particles supports a statistical analysis such as that of Bjorken and Paschos using partons. We view partons as a form of gambols.

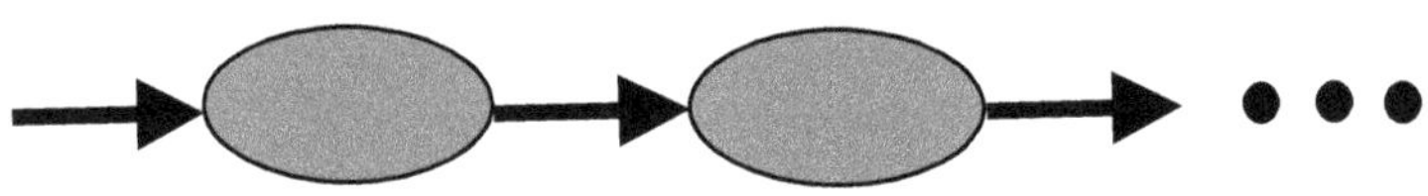

Figure 19.1. A fermion undergoing a sequence of transitions between the fermion to a set of its constituent 844242 gambols. This infinitude sequence becomes a variation on the fermion Green's when calculated in momentum space. The shaded ovel represents a distribution of 844242 gambols. One could also define interaction terms for more complex aggragates of gambols.

19.3 Lagrangian Model for a Fundamental Fermion Composed of Confined Gambols

We now define a Lagrangian model for a free PseudoQuantum fermion that may be viewed as composed of free gambol PseudoQuantum fermions. There are two apparent approaches to "embedding" gambols within fermions: one approach specifies the fermion as containing confined gambols – perhaps in a confining quantum vector field.[60]

---

[58] See also Blaha (*2005a*). The fermion triangle graph is renormalizable in this formalism.

[59] See Blaha (2023e) and references therein.

[60] We considered confinement in the 1970's due to a higher derivative $1/k^4$ vector field using a canonical PseudoQuantum formalism.

Another approach used here makes the fermion embody gambols using an interaction such as eq. 19.6 above. The fermion's Feynman propagator then acquires a term that represents states of the particle alternating with multi-gambol states. For lower energies, experiment should show the fermion in interaction. For very high energy experiment should reflect interacting multi-gambol states. This type of situation is apparent in Deep Inelastic lepton-nucleon scattering results. So we shall follow that approach here. We define Lagrangian with fermion-gambol transitions and gauge field interaction between fermions and between fermions and their gambol equivalent.[61]

$$\mathscr{L} = \overline{\psi_2}[\gamma^\mu\partial_\mu - M]\psi_2 + \overline{\psi_1}[\gamma^\mu\partial_\mu - M]\psi_2 + \overline{\psi_2}[\gamma^\mu\partial_\mu - M]\psi_1 + \overline{\psi_1}[\gamma^\mu\partial_\mu - M]\psi_1 +$$

$$+ \sum_{n=1}^{844242}[\overline{\psi_2}^{gn}[\gamma^\mu\partial_\mu - M]\psi_2^{gn} + \overline{\psi_1}^{gn}[\gamma^\mu\partial_\mu - M]\psi_2^{gn} + \overline{\psi_2}^{gn}[\gamma^\mu\partial_\mu - M]\psi_1^{gn} + \overline{\psi_1}^{gn}[\gamma^\mu\partial_\mu - M]\psi_1^{gn}] +$$

$$+ g\overline{\psi_2}\prod_{n=1}^{844242}\psi_2^{gn} + g(\prod_{n=1}^{844242}\overline{\psi_2}^{gn})\psi_2 + e\overline{\psi_2}\sum_k A_k^\mu(y)\,\gamma_\mu T_k \prod_{n=1}^{844242}\psi_2^{gn} + e\overline{\psi_2}\sum_k A_k^\mu(y)\gamma_\mu T_k\,\psi_2 +$$

$$+ e(\prod_{n=1}^{844242}\overline{\psi_2}^{gn})\sum_k A_k^\mu(y)\,\gamma_\mu T_k\,\psi_2 \qquad (19.7)$$

Eq. 19.7 has interactions between the fermion and its corresponding gambols. It also has a gauge field interaction between fermions, and between the fermion and the corresponding gambols. It is a renormalizable theory due to a Two Tier formulation.

19.4 Feynman Propagator

The denominator of the Feynman propagator in momentum space for eq. 19.7 has the form

$$p^2 - M^2 + \Sigma(p) \qquad (19.8)$$

where $\Sigma(k)$ is the term generated by the gambol interaction. It is consistent with a summation of diagrams such as those in Fig. 19.1.

At lower energies the k^2 term dominates yielding the fermion's behavior in interactions with external fields. At high energies the $\Sigma(k)$ term dominates and may have the effect of causing a behavior similar to a sum of gambol interactions with external fields.

The approach presented here avoids the need for a new mysterious gambol confining mechanism.

19.5 Gambol Statistics

The Lagrangian of eq. 19.7 is somewhat formidable with its 844242 gambols. This worry is mitigated by realizing that calculations of decays and interactions in Perturbation Theory may be well approximated by statistical calculations. We successfully used a statistical approach (Planckian distributions) in our previous books on gambols to which the reader is referred.

---

[61] The gambol interaction terms would appear to create divergences in perturbation theory. However the use of the author's Two Tier Quantization eliminates this possibility

20. Gambol Strata

Each charged fermion in a stratum has an associated gambol. In the UST of our universe there are 224 gambols per stratum with each gambol associated with a corresponding charged fermion. A charged fermion particle is "equivalent" to 844242 corresponding gambol fermions. Usually its gambol interface can be "seen" at high energy such as we saw in Deep Inelastic scattering and the other examples that we considered in previous books.

The gambol corresponding to a charged fermion has a mass equal to a 1/844242 factor of the fermion's mass. Its additive quantum numbers are 1/844242 of the corresponding fermion's additive quantum numbers. The masses of the gambol set of a fermion add to equal the fermion's mass and their additive quantum numbers add to equal the fermion's additive quantum numbers.

20.1 Gambols Within Gambols

Some or all of the gambols of a fermion can be transformed to gambols of the next lower stratum. Thus the number of gambols of a fermion ranges from 844242 to 844242^2 counting the gambols of the type of the fermion's stratum and the gambols of the equivalent fermion of the next lower stratum.

For example, the 2nd stratum u quark has 844242 second stratum gambols of the u type. If these 2nd stratum gambols are all transformed to stratum one u type gambols, then the u quark has 844242^2 u-type gambols of the 1st stratum.

In a similar manner a charged fermion of the 3rd stratum can be made equivalent to a set of 844242^3 gambols of the equivalent fermion's gambols of the 1st stratum. The growth in gambol numbers can be truly astronomical.

The possible sequences of corresponding gambols of a 3rd stratum charged fermion are:

844242 stratum 3 gambols
844242^2 stratum 2 gambols
844242^3 stratum 1 gambols

20.2 Six Connection Groups for Gambols

The transition from gambols of one stratum to another stratum can take place through Gambol Connection Groups. The Strata Connection Groups transformations for *fermions* in strata are given by Fig. 18.2. We now extend the Connection Groups to gambols as shown in Fig. 20.1.

When implementing a transition between a fermion in one stratum to the corresponding fermion in the next lower stratum, such as a transition from a stratum 3 fermion to a stratum 2 fermion the mass of the stratum 2 fermion is 1/95939 of the mass of the stratum 3 fermion. The mass of the stratum 2 gambol is 1/(95939·844242) of the mass of the stratum 3 fermion. The additive quantum numbers are 1/844242 of the additive quantum numbers of the stratum 3 fermion (and the stratum 2 fermion).

The same type of transformation applies for gambols: a gambol in one stratum transitions to the corresponding gambol in the next lower stratum. For example, consider a transition from a stratum 3 gambol to a set of 844242 stratum 2 gambols: the mass of *each* stratum 2 gambol is 1/95939 of the mass of the stratum 3 gambol. *The generated set of stratum 2 gambols has the total mass of the stratum 3 gambol. Its additive quantum number equals the corresponding stratum 2 gambols' additive quantum numbers, and equals the stratum 3 gambol additive quantum numbers. The gambol additive quantum numbers are the same for all gambol strata.*

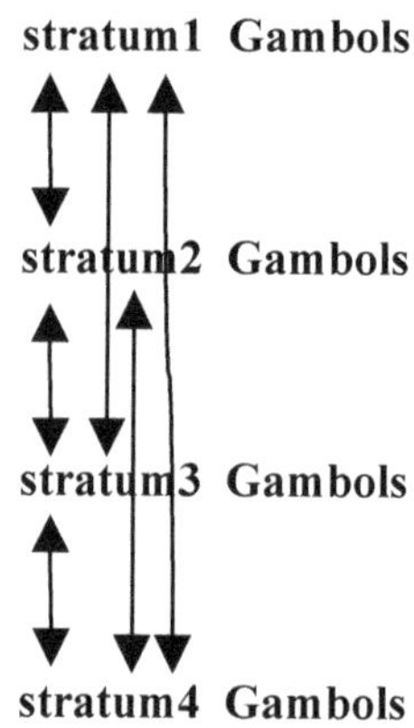

Figure 20.1. A view of the four UST strata of the Sequence Theory showing the 6 groups and interactions relating the gambols in the various strata. The interaction indicated with the arrows is between a gambol of one stratum with the corresponding gambol of another stratum. For each gambol type the masses increase by a factor of 1/95939 as one ascends the strata. The additive quantum numbers for gambols are the same for each gambol stratum. They are equal to the additive quantum numbers of the stratum 1 fermion descreased by a factor of 1/844242.

21. The Sequential vs. The Instantaneous

The description of Sequence Theory, Cosmos Theory and Gambol Theory is sequential as it must be for the description of the features and their interrelations. Consequently our development has proceeded in stages.

As we have noted the Reality of the Cosmos cannot be viewed as developing in stages in a sequential manner. In the absence of a sense of time, it must be viewed as instantaneous. The Parent universe(s) is coeval with the Cosmos. It exists outside of any time. Although it has internal time and spatial coordinates.

Subuniverses exist in time – the time of the Parent universe.

21.1 Driving Force of the Cosmos

If there were a sequential development of Cosmic Reality, then we might view the driving force of the Cosmos as generating the stages of development that we have followed in the beginning chapters of this book. However, Cosmos events *must* be viewed as necessarily instantaneous forever in the absence of time. Then one must view the driving force (Prime Mover) as a fiction that provides a justification for the sequence of stages necessary to a description.

One could make an analogy of our descriptive progression with Euclidean Geometry. Euclidean Geometry starts with primitive terms: point, line, and so on. Then it introduces postulates. And then it introduces structure: circles, rectangles, parallel lines, angles, and so on. Can we say there is a driving force for the introduction of these structures? No. One takes the introduction of these postulates and structures as features based on the logical extension of the primitives. Geometry has no driving force.

Similarly, we define a primitive: dimension. We then envelop it in features through constructions such as the fractal construction and the Master Equation. These constructions leads to the further construction of spaces and universes. We view the driving force that appears necessary to take us from construction stage to construction stage as a humanization of the process.

21.2 The Master Equation "Driving Force"

Later in chapter 22 we show the Master Equation solution embodies the evolutionary sequence of the Cosmos. Thus it gives an alternative to a simple set of driving forces for evolution. We will see it encompasses the initial mathematical point, the generation of a two dimension piecewise linear unit length, the generation of the fractal Cosmos Curve, the generation of the Cosmos spaces and their dimension arrays, the unification spaces sequence, and the set of sequences of Fig. 11.5.

The Master Equation is then the "driving force" basis of the evolution of the Cosmos conceptually. It encompasses all the stages from a mathematical point to the spaces of the Cosmos.

21.3 The Cosmos Organism

The Cosmos theories show a clear evolution from primitive beginnings to the full set of Cosmos structures and sequences, and the contents of universes that follow therefrom.

It is not a Real-world organism but it has a character that supports the notion that Physical Reality – our Cosmos Theories – have many analogous features to natural organisms – particularly the growth that evolves from its “seeds.”

It can be expected that Cosmos theory will evolve further.

22. Generation of the Initial Stages of the Cosmos

The evolution of the Cosmos from one mathematical point involved a number of ingredients: a point, the generation of a two dimension square through a fractal type of growth, the scaling of this square to an infinite two dimension space, and the introduction of the Master Equation in this space that led to the sequences and the Cosmos set of spaces.

In this chapter we shall unite that evolutionary sequence into a interrelated theoretical framework that will hopefully remove the use of "jumps" in the evolution of the Cosmos. The evolutionary sequence will follow from the Master Equation of chapter 7.

Due to its confluence of features we now take the Master Equation to be the basis of the evolution of the Cosmos:

1. The mathematical point.
2. The generation of a piecewise linear unit length.
3. The construction of the fractal Cosmos Curve.
4. The set of Cosmos spaces.
5. The set of sequences of Fig. 11.5 that specify coupling constants, fermion masses, fermion gambols, unification spaces, and so on.

.

The Master Equation smoothly combines the features of Cosmos evolution. This chapter generates most of the results of chapters 5 – 8.

22.1 Solutions of the Master Equation

In chapter 7 we began by defining a function w(r)

$$\psi(r) = r^{-1/2}\, w(r) \qquad (7.1e)$$

to obtain

$$(\partial^2/\partial r^2 + \lambda_{n_\varphi}/mr^2 + mC)w(r) = 0 \qquad (7.1g)$$

where

$$\lambda_{n_\varphi} = n_\varphi{}^2/m \qquad (7.1h)$$

and n_φ is a positive or negative integer. We now consider the solutions of eq. 7.1g for small r. The complete solution is a sum of Bessel functions.[62]

The solution for w(r) for small r is

[62] Since we can scale the r coordinate due to the scale covariant $1/r^2$ potential we can take the small r solution to be extended indefinitely.

$$w(r) \approx Br^{½} \cos((\lambda_{n\varphi} - ¼)^{½}\ln r + b) \quad (22.1)$$

implying

$$\psi(r) \approx B\cos((\lambda_{n\varphi} - ¼)^{½}\ln r + b) \quad (22.2)$$

where B and b are constants. The oscillating cosines give us the set of zeroes of the solution:

$$(\lambda_{n\varphi} - ¼)^{½}\ln r + b = \pm\pi/2, \pm 3\pi/2, \pm 5\pi\ 2, \pm 7\pi/2m\ ,,, = (n + ½)\ \pi \quad (22.3)$$
$$= (n + ½)\ \pi$$

where n is a positive or negative integer. Solving eq. 22.3 for r we find zeroes at

$$r_n = \exp([(n + ½)\ \pi - b]/(\lambda_{n\varphi} - ¼)^{½}) \quad (22.4)$$

If we set

$$b = \pi/2$$

and let n → –n, then

$$r_n = \exp((n\ \pi)/(\lambda_{ec} - ¼)^{½}) = C_{n,n\varphi} \quad (22.5)$$

where

$$C_n = \text{const}\ \exp(-n\pi/(\lambda_{n\varphi} - ¼)^{½}) \quad (7.2)$$

is the nth energy eigenvalue. Using

$$g_{n\varphi} = \exp(\pi/(\lambda_{n\varphi} - ¼)^{½}) \quad (7.2b)$$

and setting the "const" = 1 in eq, 7.2 and $g_{n\varphi=4}$ = 2 in Fig. 11.5 we find the n_φ=4 sequence gives

$$r_n = 2^n = C_{n,n\varphi=4} \quad (22.6)$$

Each r_n value is thus a Cosmos space dimension array column length. Thus the zeroes of the Master Equation solution map to Cosmos spaces dimension arrays. For positive n we obtain the HyperCosmos spaces. For negative n we find the zeroes match the Limos spaces' dimension arrays.

22.2 Cosmos Sequences

The other values of n_φ in Fig. 11.5 follow from the eq. 22.5 Master Equation solution' zeroes and give the Cosmos sequences: coupling constants, fermion masses, fermion gambols, unification spaces, and so on.

22.3 Cosmos Fractal Curve

In addition we find the r_n values for positive n are the piecewise linear curves used in the Hilbert fractal curve construction. The sum of these curves is the Hilbert fractal curve that generates the two dimension unit square. That curve is extended to a complete, continuous unit square.

The r_n zeroes of the complete set of n values: $-\infty, \ldots, 0, 1, 2, \ldots, +\infty$ gives us the full Cosmos Fractal Curve[63] which takes the $r_{-\infty}$ point of zero dimension and generates the unit dimension one piecewise linear curve, and then the two dimension unit square. The sum of r_n values for negative n is

$$\sum_{n=-1}^{-\infty} 2^n = 1 = r_0 \tag{22.7}$$

The fractal construction then uses r_n values from $n = 0$ to ∞.

Note the initial mathematical point is specified by $r_{-\infty}$. A complete progression from a point to the Cosmos spaces.

22.4 Totally Antisymmetric Tensors and n-forms[64]

Totally antisymmetric tensors may be based on n-forms.

22,4.1 n-forms

Totally antisymmetric integration elements lead to totally antisymmetric n-forms. The set of all differential forms for dimension n consists of n + 1 forms that are named 0-form, 1-form, 2-form, … , n-form. A k-form has k factors corresponding to the differential form $dx_1dx_2 \ldots dx_k$. The 0-form corresponds to an integration with no differential form factor, namely, non-integration.

The k-forms are independent and totally antisymmetric in their factors. In general the sum of the (n + 1) k-forms of dimension r

0-form, 1-forms, …. n-form

is[65]

$$2^n \tag{22.8}$$

22,4.2 Totally Antisymmetric Tensors

The above discussion of n-forms has an equivalent when one considers tensors. The tensors of an n dimension space with indices xyz… have the form:

$$T_{xyz\ldots}$$

Totally antisymmetric tensors have significant Physical applications such as the determination of the form of Dirac γ-matrices in 4 dimensions. Tensor indices range from a zero index tensor through an n index tensor. Thus there are n + 1 types of antisymmetric tensors in n dimensions. Their total number is given by eq. 22.8.

The evolution of the Cosmos from one mathematical point involved a number of ingredients: a point, the generation of a two dimension square through a fractal type of growth, the scaling of this square to an infinite two dimension space, and the

---

[63] See Blaha (2024b), (2024c) and chapter 7 of (2024d).

[64] See chapter 2 of Blaha (2024d).

[65] In an n dimension space the number of integrand factors for objects: lines, areas, and volume elements must be less then n + 1.

introduction of the Master Equation in this space that led to the Cosmos sequences and the Cosmos set of spaces.

Thus we have shown from the Master Equation solution that the evolutionary sequence of the Cosmos follows. We see it encompasses the initial mathematical point, the generation of a two dimension piecewise linear unit length, the generation of the fractal Cosmos Curve, the generation of the Cosmos spaces and their dimension arrays, the unification spaces sequence, and the set of sequences of Fig. 11.5.

The Master Equation is the basis of the evolution of the Cosmos conceptually. As we have often pointed out, the Cosmos evolution is instantaneous in the sense that there is no time coordinate and therefore timeless – amaranthine!

The Master Equation embodies the Cosmos and sets the stage for its universes.

23. Goals of Physics Theory

This world is becoming more resource constrained with time. Our ability to do advanced exploration of Physics is becoming more difficult. The difficulties are particularly evident in High Energy Physics where the needed resources are enormous and the availability of resources growing less.

In view of these issues one must ask; What goals do we wish to pursue that will have the greatest long term benefits for humanity?

It seems to this author that two goals are evident in fundamental Physics:

1. To develop an understanding of Physics at the deepest, and most rewarding, level.

2. To construct a theory of Physics that explains all in a simple, direct, succinct manner. Physics is more than a set of equations. It must embody the deepest thinking and perspectives in a compact formulation that describes everything or can grow to describe everything.

There are many roads to Physics understanding. There are many questions and possibilities that present themselves.

23.1 Example: Artificial Nuclear Matter

One example that addresses fundamental questions, and current major issues is the need for compact energy storage and generation. This author has proposed creating matter, *quarkium-9*, composed of highly compacted tritium atoms[66] that are compressed to form a 9 quark nucleus – not of sets of nucleons – protons and neutrons – but of their inner quarks.

Quarkium-9 could be created from an imploding tritium atom under a force – laser or some other electromagnetic force – that causes the quarks within the tritium to compress within the nucleus to form a quasi-stable fermion particle quarkium-9. See Fig. 23.1. The required force is extremely large and orders of magnitude beyond current implosion forces used in fusion reactor studies.

However, a form of quarkium-9 may also be created by colliding an extremely high energy proton beam on a deuterium target. The quarkium-9 would appear as a resonance in the scattering. If we use constituent quark masses as a guide with an average mass of 333 MeV/c^2 then the resonance would be at a reasonable energy of 2.997 GeV/c^2. Particles of the order of magnitude of this energy have already been found although they are not quarkium-9.

---

[66] Tritium is a bound state of 2 neutrons and a proton, which amounts to 5 d quarks and 4 u quarks. Ignoring tritium electrons it has electric charge +1. See Fig. 23.1 and chapter 8 of Blaha (2024b).

If quarkium-9 is produced then the question then becomes: What is its lifetime? Can we collect a macroscopic number of such particles in a "bottle" and then use it as an energy source – a battery?

If such a form of matter were created it would have applications in super compact batteries, energy storage for power grids, energy for vehicles ranging from automobiles to starships. A compact energy storage is the goal.

Quarkium-9 is a new form of nuclear energy.[67] We think of the nuclear fission of uranium and other radioactive fuel. We think of the fusion of hydrogen to obtain fusion energy. Quarkiun-9 is a new form of *artificial nuclear matter* that we can create that could provide an alternate (a more powerful, more compact alternative) to nuclear and fusion energy. Quarkium-9 could be a step beyond natural nuclear sources!

[67] Another possibility is a 27 quark particle: Quarkium-27. See chapter 8 of Blaha (2024b).

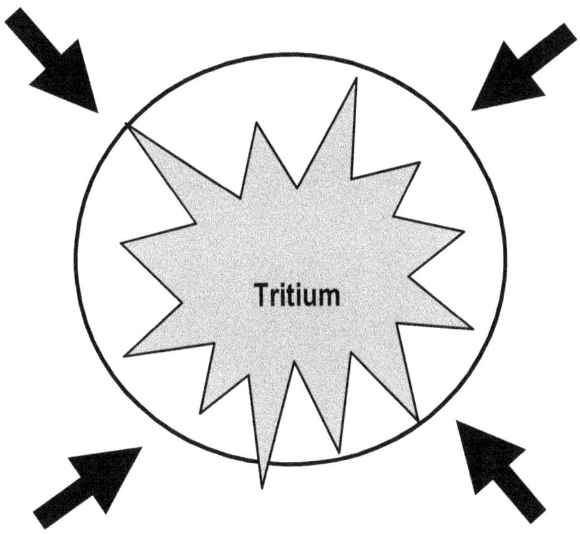

Figure 23.1. Schematic diagram of an imploding tritium atom being compressed into a quarkium-9 particle.

24. Cosmos Origin

The origin of the Physics description of the Cosmos that we have presented is clear. The description appears to be complete. The description appears to be logically sound.

If we consider the Reality of the Cosmos as far as we understand it, then our Cosmos theories appear to be a faithful description.

The question then arises: What is the origin of the Cosmos? – not in time since time does not exist until universes are created and then only within the universes.

The question has a validity since there are two situations to consider: Nothingness with no structure and no time, and our Cosmos theories, which have structure and contents and are eternal. The distinction is clear.

But the origin is elusive. The only approach to an answer that appears reasonable is:

The true origin of the Reality of the Cosmos is
beyond the scope of Physics. The true Origin is external to Physics theories.

For the Physical description of the Cosmos we believe:

The Master Equation embodies the Cosmos and sets the stage for its universes.

25. The Chain of Cosmos Development

Cosmos theory as it has evolved is the result of a significant series of stages. In this chapter we list the some of the stages with comments.

Cosmos Structure[68]	Comment
Mathematical Point	
Fractal Cosmos Curve	
Separation of Curve Line Segments	
Map to the Dimensions of Spaces	
Separation of Cosmos Two Dimension Grid into Infinitesimal dimension growth	
Map of Dimensions to totally Asymmetric Tensors	
Map of Total Numbers of these Tensors to Dimension Array Sizes	
Spaces – Space-Time and Internal Symmetry dimensions	
Dimension Structure of spaces based on quadrupling	Assumption based on antisymmetric tensor and n-form structure
Choice of certain spaces as structuring universes – 10 spaces	Arbitrary
Map of elements of each space to Internal Symmetry group dimensions	Based on Standard Model
Choice of some groups to be Connection Groups	To unite fermion sectors
Map of some elements to space-time dimensions	
Unification of Internal Symmetries and Space-time in two composite spaces	
Use of a Fundamental Reference Frame to reduce the number of dimensions to one	
Result: A comprehensive Structure without Substance	

Cosmos Universes	Comment
Substances as Energy-Matter Composites	
Universe Structure defined by a space structure	
Universes have mass-energy when created	
Universes begin as a point particle	
Universes begin by expansion (Big Bang)	
Universes are particles with a QFT with interactions	
One universe is the parent of all other universes	May be more than one parent
Universes have an internal dynamics based on space symmetries	Space structure determines dynamics

Particle Interactions	Comment
In our universe, Unified SuperStandard Theory (UST)	consistent with r = 4 space dimension array
Universes of higher dimension spaces – quadrupling versions of UST	assumption
Higher dimension particles – quadruplings of UST	assumption
Higher dimension life – possible but chemically different	assumption

---

[68] See the Appendix for details.

REFERENCES

Akhiezer, N. I., Frink, A. H. (tr), 1962, *The Calculus of Variations* (Blaisdell Publishing, New York, 1962).
Bjorken, J. D., Drell, S. D., 1964, *Relativistic Quantum Mechanics* (McGraw-Hill, New York, 1965).
Bjorken, J. D., Drell, S. D., 1965, *Relativistic Quantum Fields* (McGraw-Hill, New York, 1965).
Blaha, S., 1995, *C++ for Professional Programming* (International Thomson Publishing, Boston, 1995).
_______, 1998, *Cosmos and Consciousness* (Pingree-Hill Publishing, Auburn, NH, 1998 and 2002).
_______, 2002, *A Finite Unified Quantum Field Theory of the Elementary Particle Standard Model and Quantum Gravity Based on New Quantum Dimensions™ & a New Paradigm in the Calculus of Variations* (Pingree-Hill Publishing, Auburn, NH, 2002).
_______, 2004, *Quantum Big Bang Cosmology: Complex Space-time General Relativity, Quantum Coordinates™ Dodecahedral Universe, Inflation, and New Spin 0, ½, 1 & 2 Tachyons & Imagyons* (Pingree-Hill Publishing, Auburn, NH, 2004).
______, *2005a, Quantum Theory of the Third Kind: A New Type of Divergence-free Quantum Field Theory Supporting a Unified Standard Model of Elementary Particles and Quantum Gravity based on a New Method in the Calculus of Variations* (Pingree-Hill Publishing, Auburn, NH, 2005).
______, 2005b, *The Metatheory of Physics Theories, and the Theory of Everything as a Quantum Computer Language* (Pingree-Hill Publishing, Auburn, NH, 2005).
______, 2005c, *The Equivalence of Elementary Particle Theories and Computer Languages: Quantum Computers, Turing Machines, Standard Model, Superstring Theory, and a Proof that Gödel's Theorem Implies Nature Must Be Quantum* (Pingree-Hill Publishing, Auburn, NH, 2005).
______, 2006a, *The Foundation of the Forces of Nature* (Pingree-Hill Publishing, Auburn, NH, 2006).
______, 2006b, *A Derivation of ElectroWeak Theory based on an Extension of Special Relativity; Black Hole Tachyons; & Tachyons of Any Spin.* (Pingree-Hill Publishing, Auburn, NH, 2006).
_______, 2007a, *Physics Beyond the Light Barrier: The Source of Parity Violation, Tachyons, and A Derivation of Standard Model Features* (Pingree-Hill Publishing, Auburn, NH, 2007).
______, 2007b, *The Origin of the Standard Model: The Genesis of Four Quark and Lepton Species, Parity Violation, the ElectroWeak Sector, Color SU(3), Three Visible Generations of Fermions, and One Generation of Dark Matter with Dark Energy* (Pingree-Hill Publishing, Auburn, NH, 2007).
______, *2008a, A Direct Derivation of the Form of the Standard Model From GL(16) (Pingree-Hill Publishing, Auburn, NH, 2008).*
______, 2008b, *A Complete Derivation of the Form of the Standard Model With a New Method to Generate Particle Masses Second Edition* (Pingree-Hill Publishing, Auburn, NH, 2008)
______, 2009, *The Algebra of Thought & Reality: The Mathematical Basis for Plato's Theory of Ideas, and Reality Extended to Include A Priori Observers and Space-Time Second Edition* (Pingree-Hill Publishing, Auburn, NH, 2009).
______, 2009a, *Bright Stars, Bright Universe* (Pingree-Hill Publishing, Auburn, NH, 2008)
______, 2010a, *Operator Metaphysics: A New Metaphysics Based on a New Operator Logic and a New Quantum Operator Logic that Lead to a Mathematical Basis for Plato's Theory of Ideas and Reality* (Pingree-Hill Publishing, Auburn, NH, 2010).
______, 2010b, *The Standard Model's Form Derived from Operator Logic, Superluminal Transformations and GL(16)* (Pingree-Hill Publishing, Auburn, NH, 2010).
______, 2010c, *SuperCivilizations: Civilizations as Superorganisms* (McMann-Fisher Publishing, Auburn, NH, 2010).
______, 2011a, *21st Century Natural Philosophy Of Ultimate Physical Reality* (McMann-Fisher Publishing, Auburn, NH, 2011).
______, 2011b, *All the Universe! Faster Than Light Tachyon Quark Starships & Particle Accelerators with the LHC as a Prototype Starship Drive Scientific Edition* (Pingree-Hill Publishing, Auburn, NH, 2011).
______, 2011c, *From Asynchronous Logic to The Standard Model to Superflight to the Stars* (Blaha Research, Auburn, NH, 2011).
______, 2012a, *From Asynchronous Logic to The Standard Model to Superflight to the Stars volume 2: Superluminal CP and CPT, U(4) Complex General Relativity and The Standard Model, Complex Vierbein General Relativity, Kinetic Theory, Thermodynamics* (Blaha Research, Auburn, NH, 2012).

REFERENCES

______, 2012b, *Standard Model Symmetries, And Four And Sixteen Dimension Complex Relativity; The Origin Of Higgs Mass Terms* (Blaha Reasearch, Auburn, NH, 2012).

______, 2013a, *Multi-Stage Space Guns, Micro-Pulse Nuclear Rockets, and Faster-Than-Light Quark-Gluon Ion Drive Starships* (Blaha Research, Auburn, NH, 2013).

______, 2013b, *The Bridge to Dark Matter; A New Sibling Universe; Dark Energy; Inflatons; Quantum Big Bang; Superluminal Physics; An Extended Standard Model Based on Geometry* (Blaha Reasearch, Auburn, NH, 2013).

______, 2014a, *Universes and Megaverses: From a New Standard Model to a Physical Megaverse; The Big Bang; Our Sibling Universe's Wormhole; Origin of the Cosmological Constant, Spatial Asymmetry of the Universe, and its Web of Galaxies; A Baryonic Field between Universes and Particles; Megaverse Extended Wheeler-DeWitt Equation* (Blaha Reasearch, Auburn, NH, 2014).

______, 2014b, *All the Megaverse! Starships Exploring the Endless Universes of the Cosmos Using the Baryonic Force* (Blaha Research, Auburn, NH, 2014).

______, 2014c, *All the Megaverse! II Between Megaverse Universes: Quantum Entanglement Explained by the Megaverse Coherent Baryonic Radiation Devices – PHASERs Neutron Star Megaverse Slingshot Dynamics Spiritual and UFO Events, and the Megaverse Microscopic Entry into the Megaverse* (Blaha Research, Auburn, NH, 2014).

______, 2015a, *PHYSICS IS LOGIC PAINTED ON THE VOID: Origin of Bare Masses and The Standard Model in Logic, U(4) Origin of the Generations, Normal and Dark Baryonic Forces, Dark Matter, Dark Energy, The Big Bang, Complex General Relativity, A Megaverse of Universe Particles* (Blaha Research, Auburn, NH, 2015).

______, 2015b, *PHYSICS IS LOGIC Part II: The Theory of Everything, The Megaverse Theory of Everything, U(4)⊗U(4) Grand Unified Theory (GUT), Inertial Mass = Gravitational Mass, Unified Extended Standard Model and a New Complex General Relativity with Higgs Particles, Generation Group Higgs Particles* (Blaha Research, Auburn, NH, 2015).

____, 2015c, *The Origin of Higgs ("God") Particles and the Higgs Mechanism: Physics is Logic III, Beyond Higgs – A Revamped Theory With a Local Arrow of Time, The Theory of Everything Enhanced, Why Inertial Frames are Special, Universes of the Mind* (Blaha Research, Auburn, NH, 2015).

______, 2015d, *The Origin of the Eight Coupling Constants of The Theory of Everything: U(8) Grand Unified Theory of Everything (GUTE), S^8 Coupling Constant Symmetry, Space-Time Dependent Coupling Constants, Big Bang Vacuum Coupling Constants, Physics is Logic IV* (Blaha Research, Auburn, NH, 2015).

______, 2016a, *New Types of Dark Matter, Big Bang Equipartition, and A New U(4) Symmetry in the Theory of Everything: Equipartition Principle for Fermions, Matter is 83.33% Dark, Penetrating the Veil of the Big Bang, Explicit QFT Quark Confinement and Charmonium, Physics is Logic V* (Blaha Research, Auburn, NH, 2016).

______, 2016b, *The Periodic Table of the 192 Quarks and Leptons in The Theory of Everything: The U(4) Layer Group, Physics is Logic VI* (Blaha Research, Auburn, NH, 2016).

______, 2016c, *New Boson Quantum Field Theory, Dark Matter Dynamics, Dark Matter Fermion Layer Mixing, Genesis of Higgs Particles, New Layer Higgs Masses, Higgs Coupling Constants, Non-Abelian Higgs Gauge Fields, Physics is Logic VII* (Blaha Research, Auburn, NH, 2016).

______, 2016d, *Unification of the Strong Interactions and Gravitation: Quark Confinement Linked to Modified Short-Distance Gravity; Physics is Logic VIII* (Blaha Research, Auburn, NH, 2016).

______, 2016e, *MoND: Unification of the Strong Interactions and Gravitation II, Quark Confinement Linked to Large-Scale Gravity, Physics is Logic IX* (Blaha Research, Auburn, NH, 2016).

______, 2016f, *CQ Mechanics: A Unification of Quantum & Classical Mechanics, Quantum/Semi-Classical Entanglement, Quantum/Classical Path Integrals, Quantum/Classical Chaos* (Blaha Research, Auburn, NH, 2016).

______, 2016g, *GEMS Unified Gravity, ElectroMagnetic and Strong Interactions: Manifest Quark Confinement, A Solution for the Proton Spin Puzzle, Modified Gravity on the Galactic Scale* (Pingree Hill Publishing, Auburn, NH, 2016).

______, 2016h, *Unification of the Seven Boson Interactions based on the Riemann-Christoffel Curvature Tensor* (Pingree Hill Publishing, Auburn, NH, 2016).

______, 2017a, *Unification of the Eleven Boson Interactions based on 'Rotations of Interactions'* (Pingree Hill Publishing, Auburn, NH, 2017).

______, 2017b, *The Origin of Fermions and Bosons, and Their Unification* (Pingree Hill Publishing, Auburn, NH, 2017).

______, 2017c, *Megaverse: The Universe of Universes* (Pingree Hill Publishing, Auburn, NH, 2017).

______, 2017d, *SuperSymmetry and the Unified SuperStandard Model* (Pingree Hill Publishing, Auburn, NH, 2017).

______, 2017e, *From Qubits to the Unified SuperStandard Model with Embedded SuperStrings: A Derivation* (Pingree Hill Publishing, Auburn, NH, 2017).

______, 2017f, *The Unified SuperStandard Model in Our Universe and the Megaverse: Quarks, ... ,* (Pingree Hill Publishing, Auburn, NH, 2017).
______, 2018a, *The Unified SuperStandard Model and the Megaverse SECOND EDITION A Deeper Theory based on a New Particle Functional Space that Explicates Quantum Entanglement Spookiness (Volume 1)* (Pingree Hill Publishing, Auburn, NH, 2018).
______, 2018b, *Cosmos Creation: The Unified SuperStandard Model, Volume 2, SECOND EDITION* (Pingree Hill Publishing, Auburn, NH, 2018).
______, 2018c, *God Theory (*Pingree Hill Publishing, Auburn, NH, 2018).
______, 2018d, *Immortal Eye: God Theory: Second Edition* (Pingree Hill Publishing, Auburn, NH, 2018).
______, 2018e, *Unification of God Theory and Unified SuperStandard Model THIRD EDITION* (Pingree Hill Publishing, Auburn, NH, 2018).
______, 2019a, *Calculation of: QED α = 1/137, and Other Coupling Constants of the Unified SuperStandard Theory* (Pingree Hill Publishing, Auburn, NH, 2019).
______, 2019b, *Coupling Constants of the Unified SuperStandard Theory SECOND EDITION* (Pingree Hill Publishing, Auburn, NH, 2019).
______, 2019c, *New Hybrid Quantum Big_Bang–Megaverse_Driven Universe with a Finite Big Bang and an Increasing Hubble Constant* (Pingree Hill Publishing, Auburn, NH, 2019).
______, 2019d, *The Universe, The Electron and The Vacuum* (Pingree Hill Publishing, Auburn, NH, 2019).
______, 2019e, *Quantum Big Bang – Quantum Vacuum Universes (Particles)* (Pingree Hill Publishing, Auburn, NH, 2019).
______, 2019f, *The Exact QED Calculation of the Fine Structure Constant Implies ALL 4D Universes have the Same Physics/Life Prospects* (Pingree Hill Publishing, Auburn, NH, 2019).
______, 2019g, *Unified SuperStandard Theory and the SuperUniverse Model: The Foundation of Science* (Pingree Hill Publishing, Auburn, NH, 2019).
______, 2020a, *Quaternion Unified SuperStandard Theory (The QUeST) and Megaverse Octonion SuperStandard Theory (MOST)* (Pingree Hill Publishing, Auburn, NH, 2020).
______, 2020b, *United Universes Quaternion Universe - Octonion Megaverse* (Pingree Hill Publishing, Auburn, NH, 2020).
______, 2020c, *Unified SuperStandard Theories for Quaternion Universes & The Octonion Megaverse* (Pingree Hill Publishing, Auburn, NH, 2020).
______, 2020d, *The Essence of Eternity: Quaternion & Octonion SuperStandard Theories* (Pingree Hill Publishing, Auburn, NH, 2020).
______, 2020e, *The Essence of Eternity II* (Pingree Hill Publishing, Auburn, NH, 2020).
______, 2020f, *A Very Conscious Universe* (Pingree Hill Publishing, Auburn, NH, 2020).
______, 2020g, *Hypercomplex Universe* (Pingree Hill Publishing, Auburn, NH, 2020).
______, 2020h, *Beneath the Quaternion Universe* (Pingree Hill Publishing, Auburn, NH, 2020).
______, 2020i, *Why is the Universe Real? From Quaternion & Octonion to Real Coordinates* (Pingree Hill Publishing, Auburn, NH, 2020).
______, 2020j, *The Origin of Universes: of Quaternion Unified SuperStandard Theory (QUeST); and of the Octonion Megaverse (UTMOST)* (Pingree Hill Publishing, Auburn, NH, 2020).
______, 2020k, *The Seven Spaces of Creation: Octonion Cosmology* (Pingree Hill Publishing, Auburn, NH, 2020).
______, 2020l, *From Octonion Cosmology to the Unified SuperStandard Theory of Particles* (Pingree Hill Publishing, Auburn, NH, 2020).
______, 2021a, *Pioneering the Cosmos* (Pingree Hill Publishing, Auburn, NH, 2021).
______, 2021b, *Pioneering the Cosmos II* (Pingree Hill Publishing, Auburn, NH, 2021).
______, 2021c, *Beyond Octonion Cosmology* (Pingree Hill Publishing, Auburn, NH, 2021).
______, 2021d, *Universes are Particles* (Pingree Hill Publishing, Auburn, NH, 2021).
______, 2021e, *Octonion-like dna-based life, Universe expansion is decay, Emerging New Physics* (Pingree Hill Publishing, Auburn, NH, 2021).
______, 2021f, *The Science of Creation New Quantum Field Theory of Spaces* (Pingree Hill Publishing, Auburn, NH, 2021).
______, 2021g, *Quantum Space Theory With Application to Octonion Cosmology & Possibly To Fermionic Condensed Matter* (Pingree Hill Publishing, Auburn, NH, 2021).
______, 2021h, *21st Century Natural Philosophy of Octonion Cosmology , and Predestination, Fate, and Free Will* (Pingree Hill Publishing, Auburn, NH, 2021).
______, 2021i, *Beyond Octonion Cosmology II : Origin of the Quantum; A New Generalized Field Theory (GiFT); A Proof of the Spectrum of Universes; Atoms in Higher Universes* (Pingree Hill Publishing, Auburn, NH, 2021).

______, 2021j, *Integration of General Relativity and Quantum Theory: Octonion Cosmology, GiFT, Creation/Annihilation Spaces CASe, Reduction of Spaces to a Few Fermions and Symmetries in Fundamental Frames* (Pingree Hill Publishing, Auburn, NH, 2021).
______, 2022a, *New View of Octonion Cosmology Based on the Unification of General Relativity and Quantum Theory* (Pingree Hill Publishing, Auburn, NH, 2022).
______, 2022b, *The Dust Beneath Hypercomplex Cosmology* (Pingree Hill Publishing, Auburn, NH, 2022).
______, 2022c, *Passing Through Nature to Eternity: ProtoCosmos, HyperCosmos, Unified SuperStandard Theory* (Pingree Hill Publishing, Auburn, NH, 2022).
______, 2022d, *HyperCosmos Fractionation and Fundamental Reference Frame Based Unification: Particle Inner Space Basis of Parton and Dual Resonance Models* (Pingree Hill Publishing, Auburn, NH, 2022).
______, 2022e, *A New UniDimension ProtoCosmos and SuperString F-Theory Relation to the HyperCosmos* (Pingree Hill Publishing, Auburn, NH, 2022).
______, 2022f, *The Cosmic Panorama: ProtoCosmos, HyperCosmos, Unified SuperStandard Theory (UST) Derivation* (Pingree Hill Publishing, Auburn, NH, 2022).
_____, 2022g, *Ultimate Origin: ProtoCosmos and HyperCosmos* (Pingree Hill Publishing, Auburn, NH, 2022).
______, 2023a, *UltraUnification and the Generation of the Cosmos* (Pingree Hill Publishing, Auburn, NH, 2023).
______, 2023b, *God and and Cosmos Theory* (Pingree Hill Publishing, Auburn, NH, 2023).
______, 2023c, *A New Completely Geometric SU(8) Cosmos Theory; New PseudoFermion Fields; Fibonacci-like Dimension Arrays; Ramsey Number Approximation* (Pingree Hill Publishing, Auburn, NH, 2023).
______, 2023d, *Newton's Apple is Now the Fermion* (Pingree Hill Publishing, Auburn, NH, 2023).
______, 2023e,*Cosmos Theory: The Sub-Particle Gambol Model* (Pingree Hill Publishing, Auburn, NH, 2023).
_____, 2024a, *Cosmos-Universe-Particle-Gambol Theory* (Pingree Hill Publishing, Auburn, NH, 2024).
_____, 2024b, *Fractal Cosmos Theory* (Pingree Hill Publishing, Auburn, NH, 2024).
_____, 2024c, *Fractal Cosmic Curve: Tensor-Based CosmosTheory* (Pingree Hill Publishing, Auburn, NH, 2024).
_____, 2024d, *The Eternal Form of Cosmos Theory* (Pingree Hill Publishing, Auburn, NH, 2024).
_____, 2024e, *The Eternal Form of Cosmos Theory Third Edition* (Pingree Hill Publishing, Auburn, NH, 2024).
_____, 2024f, *Fundamental Constants of Cosmos Theory and The Standard Model* (Pingree Hill Publishing, Auburn, NH, 2024).
_____, 2024g, *Quark, Lepton, W and Z Masses of Cosmos Theory and The Standard Model* (Pingree Hill Publishing, Auburn, NH, 2024).
_____, 2024h, *Geometric Cosmos Geometric Universe* (Pingree Hill Publishing, Auburn, NH, 2024).
_____, 2024i, *Particles and Universes of Cosmos Theory* (Pingree Hill Publishing, Auburn, NH, 2024).
_____, 2024j, *Unification of the Subluminal and the Superluminal in Cosmos Theory* (Pingree Hill Publishing, Auburn, NH, 2024).
_____, 2024k, *The Dawn of Dynamic Cosmos Dimension Arrays* (Pingree Hill Publishing, Auburn, NH, 2024).
_____, 2024ℓ, *Structure and Dynamics of Cosmos Theory and the Unified SuperStandard Theory* (Pingree Hill Publishing, Auburn, NH, 2024).
_____, 2025a, *Black Holes, White Holes, and Superluminal Starship* (Pingree Hill Publishing, Auburn, NH, 2025).
_____, 2025b, *The Sequence Theory Foundation of Cosmos Theory and Gambol Theory* (Pingree Hill Publishing, Auburn, NH, 2025).
____, 2025c, *BEYOND TIME AND SPACE Fermion Strata & Consistency Conditions for Sequences, Cosmos Spaces and Gambols* (Pingree Hill Publishing, Auburn, NH, 2025).
Eddington, A. S., 1952, *The Mathematical Theory of Relativity* (Cambridge University Press, Cambridge, U.K., 1952).
Fant, Karl M., 2005, *Logically Determined Design: Clockless System Design With NULL Convention Logic* (John Wiley and Sons, Hoboken, NJ, 2005).
Feinberg, G. and Shapiro, R., 1980, *Life Beyond Earth: The Intelligent Earthlings Guide to Life in the Universe* (William Morrow and Company, New York, 1980).
Gelfand, I. M., Fomin, S. V., Silverman, R. A. (tr), 2000, *Calculus of Variations* (Dover Publications, Mineola, NY, 2000).
Giaquinta, M., Modica, G., Souchek, J., 1998, *Cartesian Coordinates in the Calculus of Variations* Volumes I and II (Springer-Verlag, New York, 1998).
Giaquinta, M., Hildebrandt, S., 1996, *Calculus of Variations* Volumes I and II (Springer-Verlag, New York, 1996).
Gradshteyn, I. S. and Ryzhik, I. M., 1965, *Table of Integrals, Series, and Products* (Academic Press, New York, 1965).
Heitler, W., 1954, *The Quantum Theory of Radiation* (Claendon Press, Oxford, UK, 1954).

REFERENCES

Huang, Kerson, 1992, *Quarks, Leptons & Gauge Fields 2nd Edition* (World Scientific Publishing Company, Singapore, 1992).
Jost, J., Li-Jost, X., 1998, *Calculus of Variations* (Cambridge University Press, New York, 1998).
Kaku, Michio, 1993, *Quantum Field Theory*, (Oxford University Press, New York, 1993).
Kirk, G. S. and Raven, J. E., 1962, *The Presocratic Philosophers* (Cambridge University Press, New York, 1962).
Landau, L. D. and Lifshitz, E. M., 1987, *Fluid Mechanics 2nd Edition*, (Pergamon Press, Elmsford, NY, 1987).
Rescher, N., 1967, *The Philosophy of Leibniz* (Prentice-Hall, Englewood Cliffs, NJ, 1967).
Riesz, Frigyes and Sz.-Nagy, Béla, 1990, *Functional Analysis* (Dover Publications, New York, 1990).
Sakurai, J. J., 1964, *Invariance Principles and Elementary Particles* (Princeton University Press, Princeton, NJ, 1964).
Weinberg, S., 1972, *Gravitation and Cosmology* (John Wiley and Sons, New York, 1972).
Weinberg, S., 1995, *The Quantum Theory of Fields Volume I* (Cambridge University Press, New York, 1995).

REFERENCES

INDEX

About the Author

Stephen Blaha is a well-known Physicist and Man of Letters with interests in Science, Society and civilization, the Arts, and Technology. He had an Alfred P. Sloan Foundation scholarship in college. He received his Ph.D. in Physics from Rockefeller University. He has served on the faculties of several major universities. He was also a Member of the Technical Staff at Bell Laboratories, a manager at the Boston Globe Newspaper, a Director at Wang Laboratories, and President of Blaha Software Inc. and of Janus Associates Inc. (NH).

Among other achievements he was a co-discoverer of the "r potential" for heavy quark binding developing the first (and still the only demonstrable) non-Aeolian gauge theory with an "r" potential; first suggested the existence of topological structures in superfluid He-3; first proposed Yang-Mills theories would appear in condensed matter phenomena with non-scalar order parameters; first developed a grammar-based formalism for quantum computers and applied it to elementary particle theories; first developed a new form of quantum field theory without divergences (thus solving a major 60 year old problem that enabled a unified theory of the Standard Model and Quantum Gravity without divergences to be developed); first developed a formulation of complex General Relativity based on analytic continuation from real space-time; first developed a generalized non-homogeneous Robertson-Walker metric that enabled a quantum theory of the Big Bang to be developed without singularities at $t = 0$; first generalized Cauchy's theorem and Gauss' theorem to complex, curved multi-dimensional spaces; received Honorable Mention in the Gravity Research Foundation Essay Competition in 1978; first developed a physically acceptable theory of faster-than-light particles; first derived a composition of extremums method in the Calculus of Variations; first quantitatively suggested that inflationary periods in the history of the universe were not needed; first proved Gödel's Theorem implies Nature must be quantum; provided a new alternative to the Higgs Mechanism, and Higgs particles, to generate masses; first showed how to resolve logical paradoxes including Gödel's Undecidability Theorem by developing Operator Logic and Quantum Operator Logic; first developed a quantitative harmonic oscillator-like model of the life cycle, and interactions, of civilizations; first showed how equations describing superorganisms also apply to civilizations. A recent book shows his theory applies successfully to the past 14 years of history and to *new* archaeological data on Andean and Mayan civilizations as well as Early Anatolian and Egyptian civilizations.

He first developed an axiomatic derivation of the form of The Standard Model from geometry – space-time properties – The Unified SuperStandard Model. It unifies all the known forces of Nature. It also has a Dark Matter sector that includes a Dark ElectroWeak sector with Dark doublets and Dark gauge interactions. It uses quantum coordinates to remove infinities that crop up in most interacting quantum field theories and additionally to remove the infinities that appear in the Big Bang and generate inflationary growth of the universe. It shows gravity has a MOND-like form without sacrificing Newton's Laws. It relates the interactions of the MOND-like sector of gravity with the r-potential of Quark Confinement. The axioms of the theory lead to the question of their origin. We suggest in the preceding edition of this book it can be attributed to an entity with God-like properties. We explore these properties in "God Theory" and show they predict that the Cosmos exists forever although individual universes (or incarnations of our universe) "come and go." Several other important results emerge from God Theory such a functionally triune God. The Unified SuperStandard Theory has many other important parts described in the Current Edition of *The Unified SuperStandard Theory* and expanded in subsequent volumes.

Blaha has had a major impact on a succession of elementary particle theories: his Ph.D. thesis (1970), and papers, showed that quantum field theory calculations to all orders in ladder approximations could not give scaling deep inelastic electron-nucleon scattering. He later showed the eigenvalue equation for the fine structure constant α in Johnson-Baker-Willey QED had a zero at $\alpha = 1$ not 1/137 by solving the Schwinger-Dyson equations to all orders in an approximation that agreed with exact results to 4th order in α thus ending interest in this theory. In 1979 at Prof. Ken Johnson's (MIT) suggestion he calculated the proton-neutron mass difference in the MIT bag model and found the result had the wrong sign reducing interest in the bag model. These results all appear in Physical Review papers. In the 2000's he repeatedly pointed out the shortcomings of SuperString theory and showed that The Standard Model's form could be derived from space-time geometry by an extension of Lorentz transformations to faster than light transformations. This deeper space-time basis greatly increases the possibility that it is part of THE fundamental theory. Recently, Blaha showed that the Weak interactions differed significantly from the Strong, electromagnetic and gravitation interactions in important respects while these interactions had similar features, and suggested that ElectroWeak theory, which is essentially a glued union of the Weak interactions and Electromagnetism, possibly modulo unknown Higgs particle features, be replaced by a unified theory of the other interactions combined with a stand-alone Weak interaction theory. Blaha also showed that, if Charmonium calculations are taken seriously, the Strong interaction coupling constant is only a factor of five larger than the electromagnetic coupling constant, and thus Strong interaction perturbation theory would make sense and yield physically meaningful results.

In graduate school (1965-71) he wrote substantial papers in elementary particles and group theory: The Inelastic E- P Structure Functions in a Gluon Model. Phys. Lett. B40:501-502,1972; Deep-Inelastic E-P Structure Functions In A Ladder Model With Spin 1/2 Nucleons, Phys.Rev. D3:510-523,1971; Continuum Contributions To The Pion Radius, Phys. Rev. 178:2167-2169,1969; Character Analysis of U(N) and SU(N), J. Math. Phys. 10, 2156 (1969); and The Calculation of the Irreducible Characters of the Symmetric Group in Terms of the Compound Characters, (Published as Blaha's Lemma in D. E. Knuth's book: *The Art of Computer Programming Vols. 1 – 4*).

In the early 1980's Blaha was also a pioneer in the development of UNIX for financial, scientific and Internet applications: benchmarked UNIX versions showing that block size was critical for UNIX performance, developing financial modeling software, starting database benchmarking comparison studies, developing Internet-like UNIX networking (1982) and developing a hybrid shell programming technique (1982) that was a precursor to the PERL programming language. He was also the manager of the AT&T ten-year future products development database. His work helped lead to commercial UNIX on computers such as Sun Micros, IBM AIX minis, and Apple computers.
In the 1980's he pioneered the development of PC Desktop Publishing on laser printers and was nominated for three "Awards for Technical Excellence" in 1987 by PC Magazine for PC software products that he designed and developed.
Recently he has developed a theory of Megaverses – actual universes of which our universe is one – with quantum particle-like properties based on the Wheeler-DeWitt equation of Quantum Gravity. He has developed a theory of a baryonic force, which had been conjectured many years ago, and estimated the strength of the force based on discrepancies in measurements of the gravitational constant G. This force, operative in D-dimensional space, can be used to escape from our universe in "uniships" which are the equivalent of the faster-than-light starships proposed in the author's earlier books. Thus travel to other universes, as well as to other stars is possible.
Blaha also considered the complexified Wheeler-DeWitt equation and showed that its limitation to real-valued coordinates and metrics generated a Cosmological Constant in the Einstein equations.
The author has also recently written a series of books on the serious problems of the United States and their solution as well as a book on the decline of Mankind that will follow from current social and genetic trends in Mankind.

In the past twenty years Dr. Blaha has written over 80 books on a wide range of topics. Some recent major works are: *From Asynchronous Logic to The Standard Model to Superflight to the Stars, All the Universe!, SuperCivilizations: Civilizations as Superorganisms, America's Future: an Islamic Surge, ISIS, al Qaeda, World Epidemics, Ukraine, Russia-China Pact, US Leadership Crisis, The Rises and Falls of Man – Destiny – 3000 AD: New Support for a Superorganism MACRO-THEORY of CIVILIZATIONS From CURRENT WORLD TRENDS and NEW Peruvian, Pre-Mayan, Mayan, Anatolian, and Early Egyptian Data, with a Projection to 3000 AD,* and *Mankind in Decline: Genetic Disasters, Human-Animal Hybrids, Overpopulation, Pollution, Global Warming, Food and Water Shortages, Desertification, Poverty, Rising Violence, Genocide, Epidemics, Wars, Leadership Failure.*

He has taught approximately 4,000 students in undergraduate, graduate, and postgraduate corporate education courses primarily in major universities, and large companies and government agencies.

He developed a quantum theory, The Unified SuperStandard Theory (UST), which describes elementary particles in detail without the difficulties of conventional quantum field theory. He found that the internal symmetries of this theory could be exactly derived from an octonion theory called QUeST. He further found that another octonion theory (UTMOST) describes the Megaverse. It can hold QUeST universes such as our own universe. It has an internal symmetry structure which is a superset of the QUeST internal symmetries.

Recently he developed Octonion Cosmology. He replaced it with HyperCosmos theory, which has significantly better features. He developed a fractionalization process for dimensions, particles and symmetry groups. He also described transformation that reduced particles and dimensions to a far more compact form. He also developed a precursor theory ProtoCosmos that leads to the HyperCosmos.

The author showed that space-time and Internal Symmetries can be unified in any of the ten HyperCosmos spaces in their associated HyperUnification spaces. The combined set of HyperUnification spaces enable all HyperCosmos dimensions to be obtained by a General Relativistic transformation from one primordial dimension in the 42 space-time dimension unified HyperUnification space.

At present the author devel;oped the Cosmos Theory that incorporates ProtoCosmos Theory, HyperCosmos Theory, Limos Theory, Second Kind HyperCosmos Theory and HyperUnification Spaces. He has introduced PseudoFermion wave functions and theory, He has related Cosmos Theory to Regge trajectories of spaces, parton theory, Veneziano amplitudes, Fibonacci numbers and Ramsey numbers. He has calculated an approximation to the difficult R(n,n) Ramsey numbers.

He has developed a Gambol Model that successfully accounts for e-p deep inelastic scattering, fundamental particle resonances, hadron scattering, and the inner structure of particles based on confinement through Casimir forces of ideal gambol gases. The Gambol Planckian Distribution was derived.

He has applied the Gambol Model to particles, universes, and the Cosmos of universes. He showed that the Cosmos may have a distribution of 23 universes corresponding to various Cosmos spaces.

Recently he showed that Cosmos Theory follows from the number of independent asymmetric tensors in a dimension r. He also showed the close parallel between the form of γ-matrices and Cosmos Theory dimension arrays. The closeness suggested that dimension arrays have the same importance as γ-matrices for fermions.

He demonstrated that the pressure of fermions within a space of dimension r balances the Casimir vacuum energy force for 18 dimensions. He showed that $2e\pi$ = 17.02 marks the critical point where pressure balances Casimir force, which implies $r = 18$ is the highest dimension Physical Cosmos space. The dimension $2e\pi$ appears to set the approximate dimension for Cosmos spaces with dimension array size $2^{r+4} \cong (17.02/8)^{r+4}. \cong (e\pi/4)^{r+4}. \cong 2.13^{r+4}$.

Now he has found the sequences of Coupling Constant values and fermion masses in the Standard Model and UST. This book unifies the Cosmos spaces spectrum, the coupling constant spectrum and the fermion mass sequences in a fundamental scaling differential equation.

www.ingramcontent.com/pod-product-compliance
Lightning Source LLC
LaVergne TN
LVHW071338210926
848486LV00003B/172